Striper :
The Super Fish

STRIPER:
THE SUPER FISH

by John Clift

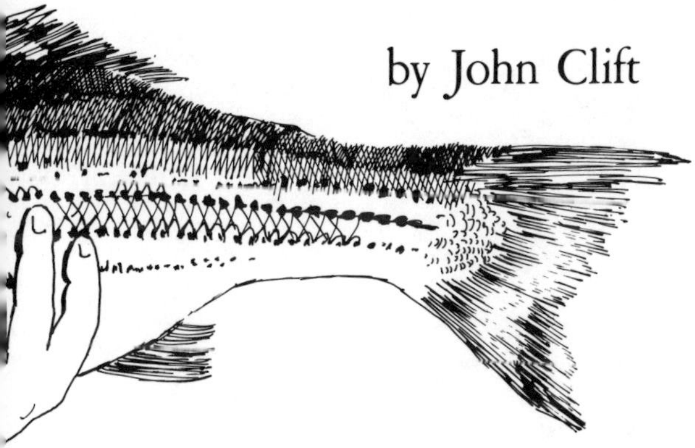

Cordovan Corporation
Houston

STRIPER: THE SUPER FISH

 Composed and printed at Houston, Texas, by the printing division of Cordovan Corporation.

Library of Congress Catalog Card Number: 75-24953
International Standard Book Number: 0-89123-009-2
Cover design and illustrations by Chuck Burkhardt

Table of Contents

Introduction

Way back in the early 1960s I attended a meeting at the Oklahoma State Park Lodge on Lake Texoma where a fisheries expert for the State of Oklahoma Department of Wildlife Conservation brought up the subject of striped bass.

I was not unfamiliar with the subject nor possibly was any other fisherman who subscribed to any national outdoor magazine. Since the mid-1950s word of this fabulous erstwhile spendid freshwater version of the saltwater striper had been making the rounds.

To a fisherman who since his childhood had dreamed Walter Mitty kind of dreams about catching the biggest black bass in the world, this was a dream come true. Here was a fish so large that even a Texan wouldn't have to brag about it. Here was a fish so big that it startled the imagination of even the most optimistic angler.

Santee Cooper was the mecca of many freshwater fishermen. Like the legions of Caesar, they came, they saw and they conquered. And they returned again and again.

But then the fisheries biologists in surrounding states wondered why they couldn't do the same thing that South Carolina was doing. They reasoned if the striper could survive and reproduce in South Carolina waters, what was wrong with North Carolina, or Tennessee, or Kentucky, or any other state?

Meanwhile, back at the ranch some 1,200 miles away, people like myself wondered why the fish wouldn't be happy in lakes such as Texoma or Keystone. After all, we had a high saline content. In fact, the late Sam Rayburn once said that if all of the salt that flowed into Lake Texoma in a single day was put end on end, it would make a tower higher than the Empire State Building. The Great Salt Plains of Oklahoma was the source of the saline content for most of the waters.

Well, as it turned out, that wasn't the most important factor. To get a natural spawn, the fish just had to have room enough to get 40 miles or so upstream where it could deposit its eggs, where the eggs would sink immediately, then rise to the surface and float downstream.

Since the mid-1950s, fishery biologists in 30 states have worked on the striped bass project. Today there are established fisheries in at least 15 states and more are being established daily. Actually, there seems to be no reason today why any state that wants the striped bass can't have them.

And isn't it wonderful to be able to live 500 or more miles from a coastal area, yet have to drive less than 100 miles in any direction to be in water where you can fish for striped bass 40 to 60 pounds?

I know it has cost money. But it cost mills compared to the billions spent in putting a man on the moon. And I'll bet today there isn't a fisherman living who has landed one of these big stripers and hasn't felt that it was the equivalent of a moon landing for him.

I hope in the following pages you'll discover how it all started, what's in store for you in your own area, some of the ways and means of catching striped bass, and what fishermen are using around the nation to land them.

John Clift
June 1975

1

I catch my first striper

Z-z-z-z-i-i-i-i-i-n-n-n-n-n-n-g-g-g-g-g!

The line on my Ambassadeur 5000 was being rapidly stripped as the big fish took off. I had been about half asleep when the fish hit. But that brought me around in a hurry.

"Fish on! Stop the engine!"

I was yelling over my shoulder as I put the pressure on the big striper. In the front end of the boat, Jodie Grigg had quickly cut off the engine. He and Earl McCullum were swiftly reeling in their lines.

Z-z-z-z-z-z-a-a-a-a-a-a-a-n-n-n-n-n-n-n-g-g-g-g-g-g-g!

Jodie had an Ambassadeur 6000. It made a more throaty sound.

"I've got one, too!" he yelled.

He didn't have to yell. I heard his reel talk first.

Some 75 yards behind the boat, my striper surfaced. It rolled in the water in a gar-like fashion.

"God, that's a big 'un!" yelled Jodie, who was busily fighting his own fish. "I'll bet it'll go 20-25 pounds!"

Then for the next several minutes we both were too busy to talk. Earl had the net handy and was swinging from one side of the boat to the other, watching the progress of the striper we each had on.

This was my first fishing trip out just for stripers. I was fishing in the big border lake that separates Oklahoma and

Texas, Lake Texoma. We were trolling in a Striper Alley about midway in front of the Denison Dam, in both Texas and Oklahoma. At the moment, I had no idea whether I had a Sooner Striper or a Lone Star State variety, if there was a difference.

Looking back, it didn't seem possible that it had been only eight years ago that the fisheries division of the Oklahoma Department of Wildlife Conservation had planted its first striper in Lake Texoma. As a matter of fact, it had been eight years, three months and one day since the first 128 adult stripers were put in this huge body of water.

But since that time, the Sooners had added 1,012,433 more stripers. And for the second time in the eight years, Oklahoma biologists were fairly confident that a natural spawn had taken place.

I was no stranger to this area. As a Boy Scout I had camped in a lush meadow hard by the Red River some 18 miles west of where I was fishing right now. I had climbed over many of its hills and walked in many of its valleys. I shot my first "wild" turkey with a .22 bolt action rifle in one of the valleys. I learned later that the farmer who missed the gobbler wasn't too wild over what had happened. But at that time, in that era, killing a wild anything was part of growing up.

I had caught fish on a willow pole with a hook baited with whatever was handy – grasshopper, grub worm, may fly. Some of the fish were scrappy black bass or battling bream. I even had taken my share of slick-skinned catfish.

And at age 11, 12 or 13, a two-pound fish was great, anything larger was an experience. And it was with a willow pole and a creek minnow that I had my greatest experience. It was a bass that must have weighed all of four pounds.

Since I had grown up here and had climbed the hills and walked in the valleys, there was a certain familiarity to the terrain that I enjoyed. Not all of it had to be out of water. I could spot that old charred oak. It had proven for all to see that lightning does strike more than once in the same spot. I was huddled under a canvas tent with seven other youngsters the first time a bolt from the blue hit the tree. That time it stripped a hunk of bark from its center all the way to the ground.

I went back in subsequent years and could recall at least one other occasion when it was hit again. So the old lightning-charred, lone oak was a landmark. And after they closed the gates on the dam that was completed in 1944 and the water started backing up and covering hundreds and hundreds of acres of this land I had once walked over, the charred oak still stood out, firm and proud in its knowledge that neither lightning nor water would destroy it.

Now it was a marker for a striper alley. There were hills that ran in several directions, and if you knew what was beneath the surface of the water, you could pretty well chart where you had to go.

And there were other markers, too. At the time we were fishing in front of the face of the dam. During the dam's construction a German prisoner-of-war camp had occupied a high point south and a little west of where we were. Taking a line on the old POW camp, you could easily line up the sloping hill now covered with 135 feet of water. And on either side, huge stripers were schooling in about 35 to 45-foot water, suspended by their own motion.

It was here that we were fishing. And it was here that Jodie and I both had a striper on.

I had been invited to come along in what Grigg, who is president of the Whopper Stopper Inc. of Sherman,Texas, had called a "lure testing trip." For almost four years he had been working to perfect his Hellbender line of lures with a new addition – a magnum Hellbender that would dig down 35 to 40 feet on its own.

He was pretty sure now he had all the kinks out. He and Earl McCollum, one of his plant officials, had been out most of the day. It was after four o'clock when I joined them.

"I think we have all of the bugs out," said Jodie. "Earl and I have been working it for the past couple of hours. We've been fishing where we felt there was no fish in order to check just how it would run. We're pretty well satisfied. Now that we know that it will run right, let's go see if it will do the most important thing – catch fish!"

We were in a bass boat equipped with a depth finder. Jodie headed north toward the middle of the dam where there was a kink in the dam as it made a tiny bend. He slowed as we neared this spot and shielded his depth finder with his hand.

"Man!" he said. "There's a lot of fish down there! And from the size of the bleeps, they gotta be stripers!"

He tossed out a red plastic marker that had 150 feet of line attached to a lead weight. The marker flipped over and over as the line played out and the lead weight sank to the bottom.

"This will mark the south end of Striper Alley," said Jodie. "Let me turn around and then we'll get some lures out. I'll troll north on what will be the east side of the point and then come back on the west side. The marker here will tell us when to turn."

Earl scored first. On our first pass over the spot where Jodie had dropped the marker his reel took off.

Z-z-z-z-i-i-i-i-n-n-n-n-g-g-g-g-g!

When the striper hit, Jodie yelled for me to come in. We both reeled in to get our lures and lines out of the way while Earl played his fish. He got it alongside the boat and it looked like a whale to me. He played it past a second time and Jodie was ready with the net. He deftly scooped it up and put it on the bottom of the boat. It jiggled happily past seven pounds on our Deliar.

Leaving Earl to take his fish off the hook and into the live box, Jodie and I got busy and put our lines back in. We were on the second pass, attacking the position from the north in a southerly manner when my first striper hit.

> **"Don't set the hook! The speed of the boat does that. Don't worry about the striper running! Your drag has been set tight enough to allow some run, but not so tight that the fish will break your line. When the striper sees the boat, it will take off again! Don't worry. Just give it some line."**

All of those directions were buzzing through my mind as I played my fish. When Jodie yelled it might go 20-25 pounds, that was just adding more whipped cream to the pudding.

"Watch out! Watch out!"

It was Earl yelling this time. My striper was heading in the direction of Jodie's line and before I knew what was happening, my striper was on his side of the boat and his was on my side.

"Here, take my rod! Hold the line tight. Hand me yours!" More orders. We switched rods – the second time. The first time I tried to hand Jodie my rod we would have really knotted the works.

"Under, not over!" he yelled. I quickly obeyed. Seconds later, we repeated the process as the two fish again changed course.

"Get the net!" It was Jodie yelling this time.

I saw why he was yelling. My fish had changed course and was heading pell-mell, full steam ahead right for the boat. I furiously worked the reel handle, trying to keep a tight line on the fast approaching fish. I was staying up. Then the fish spotted the boat.

> **"When the striper sees the boat, it will take off again. Don't worry, just give it some line."**

That message was burned indelibly into my brain. The big striper saw the boat for the first time. It broke its mad path toward the boat in a sudden halt. Then it headed for the bottom, a good 135 feet straight down.

> **"When the striper sees the boat, it will take off again. Don't worry, just give it some line."**

Man, I had the message for sure this time.

As the huge striper dove for the bottom, I reached across my Ambassadeur with my left hand and hit the button. I was going to give it plenty of line.

I almost fell out of the boat backwards when the tension left my line. I quickly flipped the rod handle to put the line back in play. It was too late. That slack was all that big baby needed to cough up that lure.

"What in hell happened?" Jodie yelled.

And I said meekly, "I just did as you said and gave him some line."

"Why do you think you had your star drag set? You don't give 'em line by hitting the panic button! You just don't try and horse him when he sees the boat the first time and makes the dive down toward the bottom!"

Jodie had been working his striper toward the boat as he alternately watched his fish and looked disgustedly at me.

Earl stood by. Jodie led the striper by once, but it lunged out a little too far for the net. But the second time it came right by the side of the boat and Earl hoisted it in.

Once off the hook, Jodie held it up. "It'll go about 10 pounds," he said. He looked at me. "Yours would have made two or three of it."

We were back in the water with our lures and Jodie was scanning the depth finder. "We're coming up on the stripers," he said. "Get ready!"

It was my turn again to yell "Fish on! Stop the motors!"

I kept my rod tip pointing straight out. Slowly I raised it, cranking swiftly as I did. I had picked up 20-30 yards of line.

Z-z-z-z-i-i-i-i-n-n-n-n-g-g-g-g!

Again the striper took off. For a moment I thought I wouldn't be able to turn it. I looked at my reel. The striper had just about all of my line out. I could see the bare spool. Sweat popped out on my forehead as I slowly raised the rod tip and tried to pick up some line.

It was slow work. I gained a little. Suddenly the striper reversed and headed for the boat. For the next couple of minutes I cranked the rod handle as fast as I ever had. Then the big striper headed for the surface and rolled.

"Keep your line tight," cautioned Earl.

It was a nice fish, but it wasn't as big as the first one I had on. Maybe that is what gave me confidence. Things suddenly got easier.

Carefully I led it toward the side of the boat. Earl was waiting with the net. He deftly slipped the net under the striper and scooped it aboard. It was a twin of the first striper.

"How'd I do boss?" I asked, proudly holding the striper aloft.

Jodie grinned.

"After what happened that first time, I think I could have caught this striper with my left hand tied behind my back!" I said loudly and proudly.

Earl rubbed his chin and then winked at Jodie.

"Did you say with your left hand tied behind you?" he asked innocently.

"Why certainly," I replied.

"It is just too bad that you didn't have it tied behind you that first time," Earl countered. "Then you wouldn't have been able to push that button and have lost the biggest dadgummed striper I've ever seen on this lake!"

2

How it all began

Man invented the steam boat, the cotton gin, the automobile and put men in space and on the moon almost before man was able to figure out how to rear saltwater stripers in freshwater. It is amazing that it wasn't necessarily man's action, but quite an accident that the striper became a freshwater fish.

In fact, the first effort to transplant stripers was not from salt to freshwater but from its native home in the Atlantic Ocean cross-country to a new home in the Pacific Ocean.

It all began back in 1879 when the fish were seined from the Navesink River in New Jersey and put in huge vats of water. From there they were hauled across the country on a freight train to San Francisco Bay.

There is a story, that obviously has grown with the telling, that crews of men went along and kept the water in the vats moving with their hands so the oxygen content would remain high.

They tell about the crew pitching off some of their weaker "passengers" every time they crossed a river, hoping they might survive in the moving water. They don't report on whether any survived. Maybe they did, maybe they didn't. But there are stripers in more than a few American rivers today. And historians duly recorded that

435 striped bass from the Atlantic Ocean survived the transcontinental journey.

By 1899, the striped bass had so increased in number that a total of 1,234,000 pounds were reported caught by commercial fishermen off the California shores. The weight of the catch pretty well leveled off after that and slowly made its way to the peak year of 1915 when 1,784,448 pounds were credited to the netters.

Along about 1915, the sports fishermen started to land some of the scrappy stripers on hook and line. It galled them to see the stripers being netted. They finally raised enough cain to get the California legislature to pass a bill

labeling the striper a sport fish and ending the assault by the play for pay boys.

By then, from the original seeding of 435 stripers, the striper was being caught from the tip of Southern California all the way up the coast to the Columbia River in Washington.

With that kind of success in transplanting a fish it is hard to understand why similar transplants in freshwater were not successful. But they were not, at least not at first, and not until the miracle of 1941.

That miracle was born of the Depression years. It came about when the Santee-Cooper Hydroelectric and Navigation Project was instigated in the early 1930s as a means of putting people to work with federal funds. When completed, the project was aimed at bringing cheap power that might help spawn some new industry for hard-hit South Carolina.

But it spawned something else. Something that was to

make more money and more friends for South Carolina than all of the power from the dams or the inland navigation, the two prime purposes of the project.

With this project, for the first time ever, large numbers of both adult stripers and fry were locked up in freshwater, away from the ocean and its salty brine. What man had been trying to do scientifically since the 1880s was done in one quick act in the closing of the dams. Once that was done, Mother Nature took over. The stripers not only survived, but they thrived in their new home.

And the big fish isn't a stay-at-home kind of fish either. The stripers love to roam. State lines are no barriers. Why, they once almost created a shooting war between fishermen from Delaware and New Jersey. One state had the fish tagged as commercial while the other as sport. It took all but the National Guard to resolve the issue before the two states finally agreed that catching a striped bass could indeed be quite a sport.

Like the cigarette ads, the striper has come a long way, baby. Well over half of the 50 states now have striper programs. And it all started in 1879 when those young stripers were seined from the Navesink River in New Jersey and transported across the continent to San Francisco Bay.

The saltwater striper record now surpasses 125 pounds. The freshwater striper record is half of that. But the records were made to be broken. And one of these days in your lifetime or mine that new record fish will be caught. And if I can hazard a guess, I would say it would come from below one of the giant dams that were built to make all of this possible in the first place.

3

Santee - Cooper striper

The mecca for all freshwater sports fishermen is 171,000 acres of water piled up behind two South Carolina dams that impound the waters of the Cooper and the Santee Rivers north of the storied old city of Charleston.

Like the women libbers of the 1970s, the two beautiful lakes created by the Santee and Pinopolis Dams retained their maiden names. Actually, the upper lake is called Lake Marion. It is the larger of the two with 110,600 acres. The lower lake is named Lake Moultrie. But to the world's freshwater striper addicts, the two are known simply as Santee-Cooper Reservoir, retaining the "maiden name" of the rivers that created them, rather than the names given after they were wedded by the dams.

Both are linked by a 7½ mile-long diversion canal. And you can bet your favorite plug that at times, the best fishing of all can be found right here in this canal.

The two lakes were named after a pair of South Carolina Revolutionary War heroes. Lake Marion got its name from General Francis Marion, better known as the "Swamp Fox." It is appropriate that the upper lake was named for him. The lake is filled with gaunt tree skeletons that jut out of the water, arms akimbo. Beneath much of the water is brush. As time goes by nature takes its toll and the trees gradually crumble, leaving hazardous hidden stumps often out of sight of a boater.

When the stripers are schooling in the early months of each year in the two reservoirs, fishermen take out in hot pursuit when they spot the diving gulls feeding on the shad the stripers have chased to the surface. And many an over-eager striper fisherman has been pulled from the water or had his rig towed in and his fishing over for the day because he didn't exercise the caution of old General Marion.

The underwater stumps are that kind of a hazard in this magnificent striper fishing lake, but then that is a tribute to the old Swamp Fox. General Marion was a planter and an Indian fighter before he joined William Moultrie's regiment at the start of the American Revolution in 1775. He earned the nickname "Swamp Fox" because of the guerrilla warfare he waged against the British. When confronted by a formidable force, he would retire to the swamps. And the British, rather than having him bottled up as they thought, found he had slipped out, shattering their lines of communication, capturing scouting and foraging parties, and in general, intimidating the Loyalists.

And today, Lake Marion can do just that to a "foreign invader" who does not use a little caution when the excitement of chasing the stripers becomes so high.

Just as Lake Marion suggests the rugged character of the man for whom it was named, so does the second half of the Santee-Cooper Reservoir. Lake Moultrie was cleared before it was flooded. Its bottom has little of the brush that fills its sister lake. Here, you can really have a ball with the schooling stripers, using much ligher tackle simply because you have more open water to play your catch.

William Moultrie was a gallant South Carolina Assemblyman before the Revolutionary War. While he prevented the British from taking Charleston in 1776, he lost Savannah in 1778, and then finally Charleston in 1780, when he was captured as his city fell. He later served as governor of South Carolina and then went to the U.S. Senate. His "Memoirs of the Revolution" in the sister states of North and South Carolina is a classic for this distinguished southern gentleman. He was indeed a gentlemen soldier and Lake Moultrie is indeed a lady-like lake with none of the hazards that you find in Lake Marion.

And it is in this reservoir, the Santee-Cooper, that the striped bass found its first permanent freshwater home. On

November 2, 1941, the impounding of water was commenced that created the lake with a total shoreline of 415 miles.

But before the dam was closed, the stripers, as had been their custom for eons, followed the Santee and Cooper Rivers in their normal spawning pattern. They came out of the sea and went far up the river to deposit their eggs, which would float down the river, hatch into fry, and the fry would then continue on down the river and into the ocean.

But the vast majority of the stripers both adult and fry were trapped behind the dam in freshwater, cut off from their customary escape route to the Atlantic Ocean. Since they had become accustomed to coming into the brackish waters of the two rivers to spawn, going far upstream to deposit their eggs, they adapted to their new home.

The advent of World War II, just a short three weeks after the gates were closed, took the headlines away from the two lakes. There wasn't a lot of time for fishing in those days. Gasoline was rationed and the 100-mile round-trip from Charleston to reach the lower end of Lake Moultrie near the Pinopolis Locks wasn't considered essential.

Thus, perhaps fate intervened once again. Because with little real fishing pressure, the stripers that were trapped behind the Santee Dam when the waters of the Cooper and Santee were impounded, grew and grew in size.

For the first half dozen years, the South Carolina Wildlife Resources Department paid little attention to the stripers. After all, fishing for the largemouth black bass, barn-door crappie and bluegill bream was excellent. Ten-pound blacks were common. Oh, an occasional striper showed up in a creel census, but at the outset not often enough for the biologists to give it more than a passing thought.

The adult stripers and fry found things a little different this year. Dams blocked their way across the two rivers and the backwash of waters was creating Lake Marion and Lake Moultrie, better known as Santee-Cooper Reservoir. Fortunately, a navigation lock at the Pinopolis Dam at Lake Moultrie permitted access to the Cooper River for some of the stripers.

Then, suddenly the stripers took over the two lakes. In 1948 and 1949 everyone who went fishing caught stripers or came back with wild tales of "stripers bigger than the boat we were in" tearing up their tackle. The South Carolina fisheries experts admitted the presence of the stripers. After all, it is pretty hard to hide a 20 or 30 or 40-pound fish.

By the late 1950s, Santee-Cooper was known worldwide for its freshwater stripers. The sport fishing industry was far more lucrative than power and navigation. It was a multimillion dollar business. And it was a business that was to attract the attention of marine biologists in most of the U.S.

At the same time, they were puzzled about the sudden appearance of stripers in such obviously bountiful numbers. As far as they and other biologists knew, weren't the striped bass anadromous? You know what I mean. Weren't they like the salmon in that they spend most of their time in salt or brackish water and return to freshwater only for spawning?

So the South Carolina Wildlife Department Fisheries biologists started a full scale investigation of this phenomena. They looked at the facts as they knew them.

Firstly, there were large numbers of young striped bass in the two lakes.

Secondly, it was almost a physical impossibility for the mature, brood stripers who had to be parents of the young, to move between the freshwater of the two lakes and the sea.

Thirdly, the only passage way in either direction was down the lower Cooper River through the Charleston Harbor and to do that, the fish had to get through the Pinopolis Lock at Moncks Corner.

Finally, to the biologists it didn't appear that such a huge migration of striped bass through the locks was possible since the locks were operated only for the occasional passage of a boat. And, the locks would have to be open for both the trip upstream and then down at the precise time when the fish were making their spawning run.

It was a most interesting situation. Sportsmen were flocking to the two lakes in daily increasing numbers and were coming back with incredulous catches and stories to match. The situation wasn't going away. And heaven forbid, that was the last thing anyone wanted. But the biologists did want an answer.

So the study was pushed. The first step was to check the lock situation. This was done by using trammel nets in a once weekly set. While in the spawning season they found stripers going upstream, they didn't find them returning in the same numbers. Thus they found that the striped bass did *not* have to return to saltwater for successful reproduction.

A tagging program was carried out with 600 stripers. It proved that the Cooper River sustained an endemic population of striped bass but that passage of striped bass through the lock was insignificant.

The biologists had some proven records to check. The South Carolina Public Service Authority, which operated the Pinopolis Dam, had a log of the total operation of the locks. They were not open often. It was inconceivable that passage of stripers at the proper time could have happened. It was doubly inconceivable that passage could have happened both coming and going.

So, there was but a single conclusion to draw. The Santee-Cooper Reservoir was definitely landlocked and the striped bass within the lakes did not have to return to the sea each year as part of their life cycle.

Logical? Yes. But at that time it was hard to believe because the conclusion was completely against all of the data that biologists had been collecting for a hundred years or more on the stripers. In fact, efforts to plant stripers in freshwater had been 100 percent unsuccessful until now.

Maybe it didn't happen at all like it appeared on the surface. Maybe what actually happened was that the striped bass were "functionally landlocked" hundreds of years earlier and their life cycle was completed in a totally freshwater habitat!

The biologists were too busy during the 1950s to worry anymore about how and why. A steady parade of fishery biologists from virtually every state in the union were wending their way to the Santee-Cooper Reservoir. They looked at the books, they looked at the lake. They took home brood fish. If the miracle happened once, couldn't it happen again? To them?

There was no shortage of fish then. The striper population of Santee-Cooper exploded, expanded, enlarged, quadrupled! Pick your own descriptive term. Whatever you select, it won't aptly describe what occurred.

Fishermen came by car, train and plane. They overran hotels, motels and fishing camps. All during the Fabulous Fifties, the average rate of catch by the anglers was *more than five fish per trip!* This kind of success on stripers was unheard of anywhere before.

Perhaps because of their own insatiable appetite for knowledge and maybe because of the pressure placed on them by their touring colleagues, South Carolina fisheries experts were learning a lot about the stripers.

They discovered after a two-year study of the stomachs of 1,634 striped bass three basic facts:

One, the striped bass of all sizes feed almost exclusively on mayfly nymphs in April and May when this food is available.

Two, they turn to clupeoid fishes – herring, gizzard shad, threadfin shad, etc. – the rest of the year.

Three, it is almost accidental for stripers to dine on other rough or on any game fish.

About the time the experts had all of this down in writing, their piscatorial Garden of Eden started to fall apart. Those huge, fat, husky stripers started losing weight. The population seemed to dwindle. Fishing fell off.

And those same hotel, motel and fishing camp owners who had been singing the praises of the South Carolina game experts to the skies, became belligerent. Fishermen became angry.

The problem was obvious. The supply of herring and shad had been depleted. There wasn't enough food in the reservoirs to sustain the striped bass.

The solution appeared hopeless. Where would the fishery experts get sufficient fish food for their stripers?

You can read the answer in the report forms on file in the office of the South Carolina Wildlife Resources Department in Columbia.

Briefly, this is what was done. Research had shown that the herring spawning run occurs in the Cooper River during March and April. So a request was made and granted to open the Pinopolis locks during these months since huge numbers of the herring concentrated below the Pinopolis Dam. This would enable the adult herring to get into the lake and successfully spawn, adding to the food supply.

Next, the state obtained an immediate supply of landlocked herring or alewives from New Jersey. This provided instant food, but there has been no indication that these fish ever became established. And, a high percentage of the herring that went through the locks and into the lakes to spawn, always die after spawning so this stock didn't provide an unlimited future supply of food for the stripers.

While waiting to see what was going to happen to their emergency food drop, the biologists pondered over what had happened. They reasoned that it was a natural cycle. At one point in the lake, the clupeoids far outnumbered the stripers. But the greedy stripers literally gorged their way into a famine situation, and this dramatically dropped the striper population, creating the situation that existed in the early 1960s. The peak and depression cycles are common in many species of game fish.

However, it seldom is indicated in such a dramatic fishing fashion as it was at Santee-Cooper. The noted reservoir actually went from a feast to a famine and many felt that the honeymoon was over. But because of the annual herring spawning run, the biologists are keeping a close watch on the clupeoids population and when necessary, they open the gates. Hopefully, this will eliminate any future drastic peaks and depressions. They now feel they have the shad and herring situation under control. And this has reaped some other benefits in that it has materially improved the fishing for bass, crappie and bluegill.

But it produced some other results with far-reaching consequences. Many lakes in South Carolina were overrun with gizzard shad. Since this fish is a vegetarian and is never caught by anglers unless accidentally snagged, it had reached as high as 70 percent of the total fish content of some lakes.

Since Santee-Cooper stripers were very much in demand by up to 30 states at this time, many of whom were just planting with no hope at all of reproduction, why couldn't South Carolina put some in its lakes that were overrun with shad?

Today, every major reservoir in South Carolina is stocked annually with striped bass fry, fingerlings or hybrids. There are no overwhelming populations of shad in any of these lakes and fishing of all kinds has improved tremendously.

By the mid-1960s South Carolina had most of the answers on the care and feeding of freshwater stripers. They knew that control of the shad was of prime consideration. They knew that there wasn't another fish to equal it among freshwater fishing addicts. And they knew if they could create a self-perpetuating control it would eliminate upkeep expenses.

The demand for Santee-Cooper striper fry and fingerlings is so great that the state has set up a priority system. States with hatchery facilities get the first choice. They work on the old Future Farmers of America and 4-H calf program basis. You know how that is. You get a heifer calf and then return the first heifer born to the program.

States with hatchery programs are given brood stock and they, in turn, return 25 percent of the hatch to Santee-Cooper. This enables dozens of states to participate in the rearing of stripers, and is one of the reasons that today more than 30 states are enjoying this magnificent freshwater striper fishing without leaving their own backyards.

Probably the key accolade to what the fishery experts have done at Santee-Cooper came through a request from the federal government relayed through the State Department.

Russia requested and has been sent striper fry. The request has been repeated, although no report yet has come from behind the Iron Curtain on how the program is doing. As one Santee-Cooper fisherman was overheard to say "we'll know how successful it is when those Russians start claiming they invented the striper."

4

Striper country

Just about every kid who ever took American history knows what the governor of North Carolina said to the governor of South Carolina. But South Carolina game and fish experts are hard pressed to know which of the other 49 states made the first request to get some of their freshwater breed of stripers.

The freshwater striped bass story started in South Carolina during the hard depression years of the 1930s. A couple of reservoirs were built to hold the Santee and Cooper Rivers with an eye to providing inexpensive electrical power to attract badly needed industry. But the industry that was attracted in the end was the free-spending tourists, who as Oklahoma's lieutenant governor George Nigh often said "are a lot easier pickin' than cotton." And the tourists came because of the striped bass.

Accidentally, South Carolina trapped these saltwater stripers that for eons had come up those two rivers from the ocean to spawn, when they closed the Santee and Pinopolis Dams. And surprisingly, the trapped stripers lived to spawn and spawn again, and thus was born the biggest freshwater boon to fishing known to mortal man. Today, Santee-Cooper is as synonymous with striped bass as ham is with eggs.

Now, the Congaree and Wateree Rivers are doing the

same thing. The Saluda River between Lake Murray and Lake Greenwood is a probable. But primarily, South Carolina has stocked all of its lakes that has a supply of shad large enough to provide food for stripers, whether or not they ever reproduce. For in the rivers and lakes where natural reproduction is taking place, South Carolina has put the basic stocking of stripers in lakes and streams of 30 states and Russia.

Naturally, not all of these states created fisheries immediately. Some may never. But the prolific striper, capable of producing 4½ million eggs from a 30 to 40-pound fish, can easily supply plenty of fish for put and take fishing through growing knowledge of hatchery programs.

ALABAMA

Joe Namath made pigskin history as quarterback for the University of Alabama, but the striper program started by the Alabama Department of Conservation and Natural Resources is making more waves with more people than Namath's heroics on the gridiron.

All impoundments of the Coosa River system have been stocked with stripers in a program that got underway in 1965. Today, Jones Bluff, Millers Ferry on the upper Alabama River, and the Guntersville and Wheeler Reservoirs on the Tennessee River have been stocked. The stocking was done at the rate of five stripers per acre on the Coosa River system to one per acre on the Tennessee River.

Excellent striper fishing exists in Alabama and it is getting better by the year. Originally, Alabama got its stripers from federal hatcheries, but in 1974 acquired the Marion National Fish Hatchery.

Barry Smith, assistant chief of the Alabama Fisheries section, said with the acquisition of the state's own hatchery it expects to raise all of its own fish and to be able to steadily increase stocking in all lakes that can adequately handle the exciting stripers.

ARIZONA

Arizona may bring to mind cactus flowers, gila monsters and sand, but if the chief of the Fisheries Division of the State Game and Fish Department has his way, visitors to this southwestern state will talk more about its striped bass fishing than its sand.

Alban R. Essback said the state striper record already is 49 pounds, but that record probably has been broken by the time you read this.

The program started in 1959 when the first of some 110,000 fingerling and yearling stripers were planted.

By 1970, Essback said a self-sustaining fishery had been established in the Colorado River, the site of the first planting. Since then Lake Mead has been stocked and some reproduction has occurred. As a result, a second fishery exists there.

Since Essback is batting 1.000 at this point, he has a striper hatchery on the drawing board that will enable him to follow the lead of other states and capture brood stripers, strip them of their eggs, and rear his own striper fry and fingerlings. When this occurs, then other lakes in the state will be stocked, even if only on a put and take basis.

ARKANSAS

That wave of excitement that swept across America in 1950 when striped bass first appeared in freshwater behind the Pinopolis and Santee Dams in South Carolina, reached the inland state of Arkansas.

Arkansas kept a careful watch on the program as it was launched by numerous states. In 1966 Arkansas was ready and moved in with both feet. Today, there is natural reproduction of stripers in the Arkansas River System.

Arkansas has stocked stripers in Greeson (Narrows), Norfolk, Beaver, Dardanelle, Bull Shoals, Ouachita, Blue Mountain, Nimrod, Maumelle and Millwood. There have been a few accidental releases in Lake Conway, Fisheries Chief William E. Keith told me. Young stripers have been found below the Ozark and Dardanelle Dams, to indicate a natural spawn in those waters.

The best fishing for striped bass has been in Lake Greeson, sometimes known as the Narrows Reservoir, or in Beaver Reservoir. Dardanelle, Ozark and the Arkansas River System produces fair to good striper fishing, but the fish are scattered and harvest is more seasonal than regular. Lake Maumelle has produced some steady catches, while good results have been obtained from just limited stocking in Norfolk and Bull Shoals Reservoirs.

It appears that Dardanelle, with its lengthy headquarters and slightly murky waters, offers the best potential for a striped bass lake in Arkansas. And a natural spawn here has really spurred the fishermen's striped bass fever.

Today, striped bass is a major part of the fisheries work in this state. Arkansas employs the shocking method to collect brood fish from its own state's tailwaters for hatching and rearing within its borders. All of this is helping push Arkansas toward the front among inland striper states.

CALIFORNIA

The Golden State has its own striper program that actually was spawned by those New Jersey stripers that survived the Transcontinental train trip in the 1870s to give the west coast a saltwater striper program comparable to that along the Atlantic.

As the Santee-Cooper picture emerged in the 1950s, California followed the experiments with keen interest and has since set up its own hatchery program of rearing stripers for both itself and other states. Many of its neighboring states have received fry, according to word they relayed when questioned about their own program.

COLORADO

Only one Colorado reservoir has been stocked with striped bass and after its initial six years, there was no confirmed natural reproduction of stripers in the lake.

Wildlife Researcher Don T. Weber said the first striped bass came from South Carolina. The fry were placed at the Las Animas Hatchery and reared to fingerlings before they were planted in the Sterling Reservoir. They were placed here since this lake was thought to have the best chance for a natural spawn.

In 1970 and 1971, striper fingerlings were obtained from California and planted in Sterling. Approximately 75,000 were planted each year.

By the end of the 1970s Colorado will add the Narrows Reservoir to its growing striper program, planting from the maturing stripers in Sterling.

Actually, Colorado obtained a half million striper fry in 1968 in exchange for an equal number of walleye. These were placed in the Wray Hatchery where all died. The second time the Las Animas hatchery was used it was more successful. Colorado also exchanged kokanee eggs for striped bass fry from California.

CONNECTICUT

Connecticut is one of the few states that had stripers and gave up after one failure. Cole W. Wilde, chief of Connecticut's Department of Environmental Protection, said that in the 1960s Connecticut stocked a number of small stripers in a pond to determine whether they could control an overabundant panfish population.

"The trial was completely unsuccessful," Wilde told me.

Of course, a decade later marine biologists are pretty well convinced that stripers dine solely on shad or may-flies. They can certainly control the shad population of any lake, but they have no taste at all for game fish.

FLORIDA

The sunshine state is all smiles with its striped bass program. And why not? It has a totally successful program that since its start in 1965 has reached the stage where the state is taking its fry from its own native brood fish.

At least seven lakes have been heavily planted with striped bass fry with positive fisheries established in two, the watchfob 100-acre Lake Hunter and the 1,000-acre Lake Julianna.

Striper fishing continues good in other Florida lakes including Talquin, Griffin, Parker, Underhill and Hollings-worth. The state had stocked more than 1,000,000 fry before it got into its own program of raising fry from native brood fish.

In Lakes Hunter and Julianna there was a drastic reduction in the shad population, which has resulted in more efforts in stocking shad than in additional stripers at this time.

GEORGIA

Georgia has one established striped bass fishery in its 15,000-acre Sinclair Reservoir where 426,021 fry showed a steady growth. But of even more importance is the fact that Georgia is now hatching its own fry from Savannah

River brood fish. All of which means that this program will be booming rapidly in the next five years.

Georgia has planted stripers now in several other lakes. Jackson Lake has had 158,792 fry planted over a five-year period with an excellent rate of growth reported.

Other Georgia lakes that have received striper fry include the 8,500-acre Blackshear Lake, which also had an adult striper planting. At Blackshear, officials estimated 70 percent of the stripers survived.

Then Lake Nottely had a striper planting that is expected to be steadily increased as the state hatches its own fry from its own Savannah River brood fish.

KANSAS

Kansas may be noted for its tall corn and pretty, bonneted maidens, but right now state fisheries experts are keeping a weather eye out on four Kansas lakes: Cheney, Glen Elder, Milford and Wilson.

These lakes are the principal water areas in this midland state. They all are federal impoundments created by damming one or more rivers and streams within the state.

"Kansas is looking for the future with its striper program," Roy E. Schoonover, chief of the fisheries division of the state, told me.

Schoonover said all four of the lakes have enough head waters to provide ample area for the stripers to go upstream for natural spawns. But Kansas, the 34th state admitted to the Union, was pretty close to the 30th state to try the striped bass from South Carolina.

Schoonover is confident enough with the early results to predict some great things not only in the four lakes where the initial plantings took place, but eventually in other fishing lakes in the state.

KENTUCKY

Kentucky was one of the early bird states along with Arkansas and Tennessee to start transporting stripers into inland waters. The first take occurred in 1957 when Kentucky sent men to South Carolina. There, they fished in the Santee-Cooper lakes or tailwaters.

They bought their out-of-state fishing licenses just like the next man, only they were seeking live stripers. When they caught a striper, it was put in a live box instead of an ice chest. They got brood females and male stripers. It was a long lonely road to hoe, but that is how it was done in Kentucky through 1962.

These adult stripers were brought back to Kentucky. Only 12 stood the truck ride home the first year and they were dumped into Lake Cumberland. The largest was only 12 pounds. The next year double that number went into Cumberland, along with 95 in Herrington and 215 in Kentucky Lake.

And so it went, with Kentucky Lake getting only 43 in 1960 while Cumberland had 119 and Herrington 69. Then Kentucky became involved in the fry effort and from 433 stripers in Lake Cumberland in 1962, leaped to 540,000 fry the next year.

Since then Barkley Lake has had three million fry in a single year, Dewey Lake a half million and Green River Reservoir 200,000. And the number of fingerlings added go into the hundreds of thousands. The stripers have come from Tennessee and from the Corning, Arkansas, federal hatchery, in addition to the old hook and line method.

The tailwaters of Kentucky and Barkley Reservoirs are considered fisheries today as are Lakes Herrington and Cumberland. At last report the state record by a rod and reel angler was 44 pounds, although commercial fishermen have pulled them out above 50 pounds.

Charles C. Bowers Jr., director of the Division of Fisheries, sees some spectacular striper fishing ahead for Kentucky.

LOUISIANA

Those cagy Cajuns don't make many mistakes. Right now they are batting 1.000 in the striped bass department, having planted striper fry in two lakes, Toledo Bend, which separates Louisiana and Texas, and D'Arbonne, a 15,000-acre lake. Fisheries have been established in both lakes.

In the tailwaters behind the dams at both reservoirs, Louisiana is taking brood fish through shocking methods and hatching their own eggs, rearing their own fry and fingerlings, and are making the striped bass rapidly into the state's most exciting fish.

There has been a slight shad reduction in both lakes

since the fisheries were established, but because the fisheries know what is necessary to keep their shad population up, it shouldn't affect their program in the future.

Catches of sizeable 20 and 30-pound stripers are becoming common in the two lakes, which indicates the total success of the program, and emphasizes that this can easily become the state's No. 1 game fish, although it will never replace crayfish in the hearts of the true Cajuns.

Louisiana has been stocking striped bass since 1965, Arthur M. Williams, fisheries biologist for the Louisiana Wild Life and Fisheries Commission, told me. In addition to the two major lakes, stocking also was done in Lake Bruin, as well as in the Calcasieu, Pearl and Tchefuncte Rivers.

MAINE

Freshwater striper fishing in Maine is somewhat of a problem. It is probably a good thing for the overall program that the old election cliche that "as Maine goes so goes the nation" is not true to the striper program in America.

Lewis N. Flagg, marine resources scientist for the Department of Maine Resources, admitted that the striped bass in Maine is exploited primarily in surf and estuarine areas.

All of the major rivers in Maine are impassable to striped bass because of head of tide or near head of tide dams which actually exterminated native striper runs in the late 1830s.

The lower estuary of the Kennebec River is a large freshwater basin which supports good striper sport fishing in the spring and early summer.

MASSACHUSETTS

The importance of striped bass was recognized as early as 1639 when the General Court of the Massachusetts Bay Colony ordered striped bass protected for game purposes.

While the Bay State has no known freshwater fisheries for striped bass, the striper still is caught in bountiful numbers as a saltwater fish off the coast and in the state's rivers.

MISSISSIPPI

The current state record for striped bass in Mississippi had climbed to 20 pounds as 1974 ran its course, but in the tailwaters of the Ross R. Barnett Reservoir, stripers up to 25 pounds had been taken. In fact, Harry Barkley, Fisheries Coordinator for the Mississippi Game and Fish Commission, said flatly that "the way stripers are growing in Mississippi, our record will be changing constantly as larger and larger striped bass are taken."

It was in 1973 that Barkley admitted a fishery had been established in the Barnett Reservoir. Planting had been going on for five years at that time, with the 33,000-acre lake stocked with 396,769 striped bass fry. Barkley said this lake offered Mississippi anglers the best opportunity for a successful trip. The Barnett Reservoir, incidentally, is near Jackson.

In the 30,000-acre Sardis Reservoir, located near Sardis, 89,500 striped bass fry were planted during the same five-year period. Additional planting probably will be necessary to improve fishing for stripers here.

Two other smaller lakes have been planted with striped bass fry. They included the Okatibbee Reservoir, a 4,000-acre lake near Meridan, and Lake Bogue Homa, which has 1,200 acres and is near Laurel. With one fishery assured in the state, that will be the basis for future stockings aimed at providing at least put and take fishing in the future.

MISSOURI

The first introduction of striped bass to the Show Me State occurred in May of 1966. And once again it was Consolidated Edison who put the half million fry in the Taum Sauk lake where Con Ed's Uniton Electric Power Plant was in operation. The purpose was to determine the survival of fry in a pump-back type hydroelectric installation.

When Union Electric launched its program the state got interested. Stanley M. Michaelson, fishery biologist, figured they ought to be an excellent fish for Lake of the Ozarks. Union Electric's Grand Glaize Hatchery had been rearing bass for the lake for 35 years, so that hatchery took over the striped bass operation.

Since the program was launched in 1967, a total of 10,450,000 fry were obtained through 1974 with 817,750 fingerlings produced. Production of striped bass for the most part has been carried out at the Glaize Hatchery.

Fish are being caught in the Lake of the Ozarks as well as in the tailwater area below the Bagnell Dam. Missouri has been involved in a trial and error effort and by the mid-1970s was starting to show some solid gains. With all indications that a fishery was being established, Michaelson moved swiftly ahead to use the fishery as a management tool to make this new exciting trophy fish an intregal part of Missouri's fishery program.

NEBRASKA

The Big Red of Nebraska grid fans is the fantastically successful University of Nebraska football team. But if Bob Thomas, chief of the Aquatic Wildlife Division of the Cornhusker state has his way, Nebraska fishermen soon will be cheering equally as loud over the state's striped bass program.

Nebraska, like its sister state of Kansas, is a late starter in this piscatorial sweepstakes. But Thomas said catchable populations of striper already exist in McConaughy Reservoir.

In addition, Nebraska has stocked striper fingerlings in Harlan County Lake, Swanson Lake, Southerland Lake, Malone Reservoir, Jeffrey Reservoir, Johnson Lake, Red Willow Lake and Medicine Creek Lake.

NEW HAMPSHIRE

New Hampshire, like Connecticut, has made only one pass at the striped bass, and like Connecticut, it met with failure.

Richard G. Seamans Jr., chief of inland and marine fisheries division of the State Fish and Game Department, said "we attempted to develop a freshwater strain of striped bass, but were unsuccessful during our one shot."

And then Seamans more or less explained why they didn't pursue the subject any further when he simply told me that "we have an excellent saltwater fishery."

And since New Hampshire is a small state and no one is

really too far from the coast, the saltwater striper fishing is more than adequate.

NEW YORK

New York has a striped bass fishery, but it has been created by stripers that come up stream from the Chesapeake Bay into the Hudson River. The state of New York has accepted this phenomena and fishermen in the state are enjoying striper fishing in an area on the Hudson from Newburgh north.

The Hudson River is tidal all the way to the federal dam at Troy, and the area from Newburgh to Troy provides excellent striper fishing. And it is getting even better, thanks to a program launched by Consolidated Edison.

Con Ed was hounded because its nuclear plant was taking a toll of brood stripers as they went upstream past its plant, and then some of the still hatching eggs were lost as they floated downstream past the plant. But a hatchery was designed across from the Con Ed plant by Texas Instruments Ecological Services Inc., that harvests the spawning fish and returns many more fingerlings to the river through a man-made hatch than probably would survive a natural spawn.

NORTH CAROLINA

Five rivers and eight reservoirs or lakes in North Carolina have been stocked with freshwater stripers. The resultant growth of the striper program has been exceeded only by South Carolina. In fact, it is the Roanoke River hatchery that is providing a large percentage of the brood stripers used in rearing fry for the other 29 states involved in the program.

North Carolina has a hatchery on the banks of the Roanoke River at Weldon. One of the early problems in obtaining freshwater striper fry was in obtaining the ripe female stripers. When the female striper's eggs become ripe, she does not feed. What's more, she completes her spawning within an hour after the eggs become ripe. This only adds to the problem of finding enough naturally ripe females to operate a hatchery.

North Carolina solved all this by allowing fishermen to

use bow nets. So anglers drifted the river by day and night and dipped into the spawning "fights." If a ripe female was taken, she was rushed to the hatchery right on the bank of the river. There, the eggs are removed by hatchery personnel, fertilized and placed in incubator jars. And the fisherman not only gets his fish back, but is paid off at the rate of twenty dollars per million eggs. Since a 40-pound striper can produce 4½ million eggs, this can be a very profitable hobby during the spawning season.

In addition to the Roanoke River, stripers have been planted in the Tar, Neuse, Cape Fear and New Rivers, according to William B. Smith, assistant chief of the Tarheel division of inland fisheries. Natural spawns have been experienced in all five of the rivers.

Kerr Reservoir, located 40 miles above Weldon on the Roanoke River, was one of the first lakes after Santee-Cooper to show a natural spawn. This occurred in 1956, three years after the initial planting. Since that was so soon after the initial planting, it was felt that Kerr might have had the same thing happen to it that Santee-Cooper had in that some striped bass were trapped in the lake at the time of impoundment. At any rate, it is one of the more productive lakes.

Other North Carolina lakes that have adapted to striped bass include the Gason and Roanoke Rapids Reservoirs, along with Lakes Michie, Norman, High Rock, Badin and Hickory.

OKLAHOMA

The most spectacular fishing in the Sooner State is found these days in the Keystone Reservoir and in its Arkansas River tailwaters along with the two-state border of Lake Texoma and the Red River tailwaters below the Denison Dam.

Keystone has been the most successful, with striped bass taken in the 30-pound class regularly, and even larger fish in the tailwaters. Because of the series of navigation locks on the Arkansas, the entire system has been subject to natural spawns.

Texoma had two positive natural spawns by the end of 1974 and one probable in the decade since the initial planting of adult stripers. And well over one million stripers have been put into the lake in addition to the natural spawn.

Fingerlings have been planted in Salt Plains, Canton and striper fry in the Grand and Foss Reservoirs. Thanks to the Arkansas System, stripers now are in Lake Eufaula and Webber Falls. There was a planting of fingerlings and adults made in Lake Murray, but no appreciable results have been noted.

During certain times of the year, the area directly below Lake Keystone becomes a spectacular spot for striper fishing. Boaters push off in John boats and make their way right up to a steel cable that crosses the river to mark the off limits area directly below the generating system at the dam. Here, literally tons of huge stripers are caught.

RHODE ISLAND

In 1959 Rhode Island stocked striped bass in some of the state's fresh waters. Fisheries Chief John M. Cronan said some of the fish survived. But he said it was rather obvious that no natural reproduction was taking place.

"Rhode Island feels that a freshwater striped bass program would not be successful in any freshwater body that did not have a significantly large tributary system," Cronan said.

And since Rhode Island is the watchfob of all 50 of the states in these United States, what Cronan is saying is simply that there isn't room enough for such a big fish in such a little state. But Rhode Island has tried.

TENNESSEE

One of the most progressive striped bass programs outside of the Carolinas is carried on in the Smoky Mountains of Tennessee. From 1958 to 1963 Tennessee cooperated with Kentucky in placing 2,000 adult and yearling stripers in the 158,000-acre Kentucky lake. By the spring of 1963, Tennessee biologists were convinced that if they were going to have striped bass it would have to be by some other method.

In 1964 Tennessee started its own hatchery for striper eggs, buying their initial eggs from commercial fishermen along the Roanoke River in North Carolina. By then South Carolina had started to use hormone injections in brood stripers to induce spawning so the eggs could be stripped.

This really put Tennessee in business and until 1971, it continued to be dependent upon outside sources for fry. But by 1971, the state had its own supply of brood fish and since then has had an unlimited supply to offer its various lakes and streams. The eggs are hatched and fry and fingerlings reared at the Morristown and Springfield hatcheries.

Striped bass are now being caught in Kentucky Lake, Cherokee, Norris, Percy Priest, Boone Lake, Wats Bar Reservoir and Fort Loudoun. Both the Tennessee and Cumberland Rivers have been stocked along with Barkley Lake.

Stripers up to 50 pounds are caught with regularity in some of the larger Tennessee lakes, and there is little doubt but that the advent of the striper to freshwaters of this state has been one of the most exciting happenings for fishermen since the invention of the push button reel.

UTAH

Just because Brigham Young didn't have the chance to catch a striped bass in Utah doesn't mean that today's Utah anglers can't. The year 1974 saw the first planting of striped bass fingerlings in the state's history in Lake Powell.

Project Leader Bruce E. May told me annual stockings with fish reared at Utah's experimental culture station will be carried out.

VIRGINIA

The Kerr Reservoir divides North Carolina and Virginia and is located partially in both states. It is one of the more successful striped bass lakes in the nation, topped only by the South Carolina twins, Lakes Marion and Moultrie that back up from the Santee and Cooper Rivers.

Virginia Fisheries biologist William E. Neal says that Kerr is the only lake in Virginia that contains an established striped bass population. From this lake, however, Virginia has planted stripers in 17 Virginia lakes on a "put, grow and take" basis. The lakes range in size from a mill-pond of 37 acres to Smith Mountain Lake at 20,000 surface acres.

The most spectacular results have occurred in the largest of the lakes, the 20,000-acre Smith Mountain Lake. This is a mountain lake with steep shorelines and water up to 250 feet deep. There is little or no shoreline cover and a large gizzard shad population.

"Our only problem here," admitted Neal, "is simply that fishermen want us to put even more stripers in the lake. They would like to see 400,000 or more put in annually.

Results in Clayton, 4,500 acres; Chesdin, 3,300 acres; and the smaller 400-acre Smith lake have been poor. Results were fair at the 630-acre Carvin's Cove, and results ranged from fair to good at the 900-acre Prince lake, the 800-acre Meade lake, and at the 300-acre Harwood Mills lake.

WEST VIRGINIA

The first seven years proved to be all bad luck for West Virginia in its efforts to create a striped bass fishery. But West Virginians are a hardy lot. They are not about to give up on something they see becoming so successful elsewhere.

By the mid-1970s, West Virginia fish biologist Bernard F. Dowler told me he was planting fingerlings at the rate of 100,000 annually.

"That's not many when you have 45,000 acres of water, but we remain optimistic that some degree success will be realized," he said.

In its first seven years, West Virginia planted bass fry, striped bass fingerlings and adult stripers. Many were put in the Ohio River. Fishermen have reported only scattered catches and at this writing the record striper has been measured in inches, not pounds. Hardly enough to put on the books.

CONCLUSION

South Carolina's highly talented Chief of the Dennis Wildlife Center, Jack D. Bayless, told me the biggest majority of states that have tried to get striped bass programs started, have been successful. Only North and South Dakota could be classified as unsuccessful. Some, by the mid-1970s were not completely out of the woods, but Bayless felt it could work. Some have not reported fully to Bayless and our reports probably will predate his.

And Bayless said that South Carolina's famed Santee-Cooper stripers are now planted in Puerto Rica and South Africa along with Russia. While no reports have come back from the latest two, Bayless said the striped bass "invented in the freshwaters of the Santee and Cooper Rivers after they were dammed, have been reared successfully behind the Iron Curtain."

5

A shocking fish story

Kim Erickson stood spraddle-legged in the bow of the flat-bottomed aluminum John boat. Jutting out in front of the bow was a three-foot boom. An electrode dangled from the boom, creating tiny wakes not unlike the sparks tossed off by a chain hanging under a truck that bounced off the highway from time to time.

On the stern end of the boat, about halfway down from midship, a generator hummed angrily. In the bow end of the boat just far enough back to give Erickson balancing leg room, was a tub filled three-fourths with water. Electric cables snaked their way around the tub and ran toward the boom. Its purpose was to shock any large female or male stripers in the tailwaters of the Arkansas River below Keystone Reservoir in Central Oklahoma.

The purpose was plain. Erickson was a biologist for the Oklahoma Department of Wildlife Conservation. He had a crew on hand seeking to net striped bass females filled with eggs. It wasn't an easy job. But it had its advantages.

One 20-pound female could produce one million eggs. A 30-pound female could return three to four million eggs. You didn't need a pocket calculator to know that if you could land this kind of a brood striper, you could produce enough stripers to really stock the lakes in your state.

At this point, Erickson's job was to try and land the brood stripers. He had earlier visited Santee-Cooper's operation in South Carolina. They had both a small 14-foot John Boat and then larger boats with larger generators that carried as many as three electrodes dangling out in front of the boat.

Oklahoma had both, but Erickson was on his own mission in the smaller craft on this blustery April day.

One of Erickson's crew was manning the engine as Erickson stood in the bow and directed the pattern.

Actually what Erickson was doing has been done illegally by anglers for generations. It is referred to as "telephoning" when performed by poachers who move into densely populated area of fish, drop their electrodes over the side, perform a quick shock and then scoop up the fish.

Working below the dam with the discharge on to bring the fish swimming upstream it was a little different. The shocker boat is used for a variety of reasons but on this spring afternoon it was being used solely to land pregnant stripers.

"Keep it closer to the rip-rap!" yelled back Erickson to his pilot. "No, not that way, but keep it parallel. Don't get too close to the tubes, we haven't got enough motor on to keep us out!" he warned.

Erickson couldn't forget the time just a season earlier when a similarly equipped boat with two men aboard had its motor conk out and it was sucked into the tubes when working about the dam. And the boat that went in after it to try and save the men, even with a larger motor, also couldn't fight it.

Both boats and six men went through the tubes. Miraculously only one was lost, and the one who didn't make it had his life jacket ripped off. So that was warning enough, and Erickson's pilot cut across the river and then in a tacking fashion worked back and forth from bank to bank.

A flash of silver caught Erickson's eye and he yelled instructions. The boat responded and his longhandled net went out and with a quick, circular motion he hit the water. It was a near miss. But it turned out to be only an addled shad.

On the bank, another crew was standing by waiting to take any stripers caught. They would perform the necessary work to get the fish in shape for their ride to the nearest fish hatchery at Bryon, some 150 miles away.

Adult striped bass are placed in a wooden agitator tank

Kim Erickson, Oklahoma assistant biologist, reaches out for an addled fish during a shocking operation in the Santee River.

equipped with supplemental oxygen. Ice is added to maintain the water temperature below 71 degrees. Quinaldine and sodium chloride are added to the agitator unit.

Erickson had a 230-volt AC single-phase generator that supplied the current for the electrode. The electrode was made of 25-foot by three-quarter inch stainless steel cable. This electrofishing is conducted during the day and nighttime hours during the spawning season of the stripers.

During the spring use of the shocker boats, just about every member of the hatchery crews get their turn in the boats. The work is alternated so that they all get a crack at every part of it.

"You know," mused Erickson during a period when he was dumping a striper out of his tub to the waiting agitator tank on shore, "it is hard to explain but there is a certain thrill when you see a fish surface and you go at it with the net. It is akin to what I feel when I'm using a rod and reel and get a shoulder-numbing strike. Until the fish surfaces or I land it, I wonder what it is. It is the same here."

When this program was first launched in South Carolina, during the first season of operation more than 900 adult female striped bass were captured and examined. Not one was in the correct state of ripeness to permit use of the eggs. In fact, during the first several years, thousands of female stripers were examined in this manner and only one was found to be "naturally ripe" and ready to drop her eggs.

South Carolina had started the project because it had

Once netted, Kim brings the striper to the boat during the shocking below Keystone Reservoir.

worked successfully in North Carolina. But in a completely different manner. North Carolina biologists had observed that during the spawning run, hundreds of fishermen lined the banks of the Roanoke River. They often landed females that were actually spilling their eggs.

So they started "operation striper reward." They notified the fishermen that when they caught a "ripe" female, they would pay if that fish was rushed to their nearby hatchery so they could be striped. Once the eggs were taken then the angler not only got his reward – depending upon the number of eggs – but he also was given his fish back.

As a result, during spawning season both commercial and sport fishermen lined the banks of the river adjacent to the North Carolina hatchery. Using bow nets, they worked hard at getting the "ripe" females. It has been a successful program here but not in other states who didn't have the advantage of a dockside hatchery.

So the electrofishing was created. The electricity does not kill the fish. It merely stuns them. They are returned to the water unharmed if they are not the species sought. Actually, this type fishing has been done by fish and game departments for years to make creel censuses and to check on other aspects of fish planted in lakes and streams.

Any person who comes up on a shocker boat for the first time may get quite a shock. In fact, as he sees the addled fish come gasping to the surface looking for air, he may want to call the nearest game warden and put a halt

It's a female striper, so Kim starts out of the boat with it.

to this unsportsmanlike fishing technique. But the fisheries people usually have a game warden or two on hand. They not only help with the actual work, but they can keep the public informed of what is going on without a group of fishermen pushing a panic button.

Since the prime target of the biologists is the 20-pound or larger female striper with the million or more eggs, the shocking seldom affects the smaller fish. The reason is that the larger the fish, the greater the expanse of body exposed to the shock.

As Erickson and his cohorts have learned, you have to be light on your feet and deft with the net to score. As the

fish surface, the boat pilot is taking off with sudden bursts of speed in all directions. More than once the man in the bow has taken an unexpected dip in the swirling, cold waters.

But it is all for the sake of improving fishing for man. In the high excitement of the chase, at times submerged objects are hit, the boat is accidentally beached, or even two boats run together.

Actually no more than from a dozen to perhaps twenty females are needed during a spawning season – if they can be successfully transferred to the hatchery in time.

Once the female striped bass is landed, it is then rushed ashore for the land crew to work on. The female is first injected with gonadotropic hormone at the capture site. Then an antibiotic, oxytetracycline, is injected to reduce mortality. Either an injection or a spray tranquilizer is given to keep the fish from panicking in its small quarters during its ride to the hatchery.

Six females and three males were caught during the period that Erickson's crew was working. But only one of the six females was at the stage where the gonadotropic hormone was able to successfully induce premature ovulating to salvage the eggs. Approximately 750,000 eggs were collected from this female and the eggs fertilized by one of the male stripers. Hatching was completed within 59 hours and 500,000 fry were produced. Some 400,000 fry survived to 12 days and were stocked in culture ponds.

Hatchery Superintendent Jack Harper and his staff discovered that once the hormone is injected, if the female is going to spawn she will do so within 36 to 48 hours. Those that occasionally spawn earlier have a much lower rate of developed eggs and there is a low rate of recovery.

During the early stages of this program, the only way the fish hatchery biologists had for removing the eggs was by taking each ripe female and slitting open her belly to remove the eggs. Thus, the survival rate of the mama striper was nil. The reason the eggs were taken this way was that in spite of the injection of relaxing tranquilizers, mama striper was every bit as keyed up at hatching time as her human counterpart.

e female striper is handed over to a waiting game ranger and Kim and his 'tner go back to work.

"It is quite an emotional trauma to lay a million eggs more or less," said Harper dryly.

One of the game rangers who had been assisting in this program had an idea. "What you wanna do is relax this broad – that is, this female striper, ain't it?" he asked.

"Now the way I see it, that's what you're trying to get this old sow striper to do. To excrete them eggs. So, why not just knock her in the head? You ain't gonna kill her. Look."

And the game warden picked up a hammer lying on a shelf and rapped the striper smartly across the side of the head. The flopping stopped immediately.

"So, if I killed her, so what?" asked the ranger. "The way you was doing it, you got the same result."

It did work. And the first fish so treated delivered her eggs, the male was milked of his sperm, the eggs were fertilized and the hatch successful.

But like everything else, the action of the game ranger resulted in some more experimenting with the tranquilizing of the fish. Today, the "hit-in-the-head" method has been succeeded by an equally successful use of an anesthetic. But the end result is the same. Now, the females have a better than even chance of surviving their hatchery spawn and being returned to a normal life in their own surroundings where they can be caught as any self-respecting striper wants to be caught – on the end of a hook and line after a tremendous struggle with a two-legged man.

6

The eggs and I

When the budget committee of the South Carolina legislature was reviewing the budget of the Game and Fish Department of the state, one of the eager-beaver committee members raised not only a question, but the eyebrows of the rest of the committee.

"What does the hatchery department of the game and fish department need with silk stockings?" the legislator asked.

In one way it is too bad that the silk stockings didn't work. At least the capital outlay was considerably less than what proved to be the ultimate solution.

The silk stockings were being used as a screen over the overflow valve on the hatchery tank where the baby stripers were put immediately after the eggs hatched. This was to keep them from washing down the drain.

But hatchery workers found they had the same problem with silk stockings that so many women complain about. They got runners in their hose, too.

So another method had to be devised. The final solution involved the use of brass sheeting – and everyone has a pretty good idea of how much brass costs – especially brass that contains 1,109 perforations per square

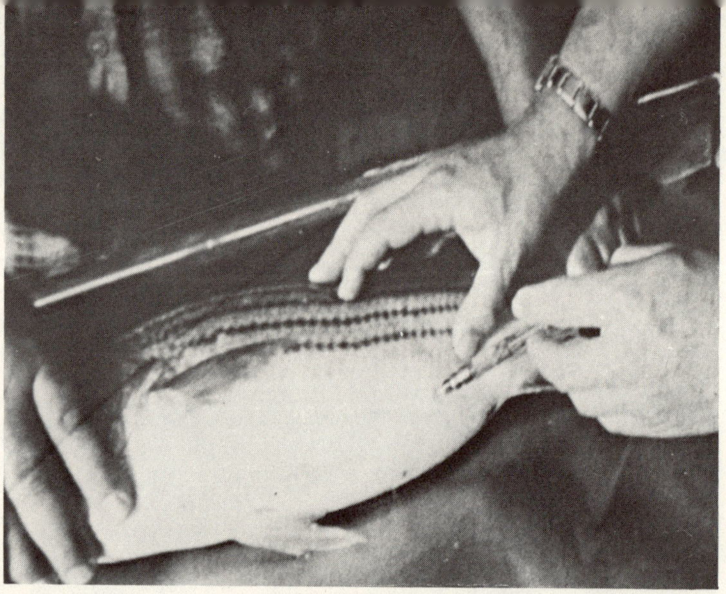

A striper is given a tranquilizer before the eggs are taken.

inch. That is such a fine mesh that it takes a really bright day to be able to hold it up to the sun and actually see that it is perforated.

The hatchery operation is the most important phase of the striped bass program. Because without the baby stripers to ultimately plant in the lakes and streams of this country, one of the most exciting species of fish wouldn't be available in freshwater lakes. Only in the sea.

States that have been able to take brood stripers and artifically hatch their eggs start their season around April 1. It lasts until about May 15, although the dates may vary from state to state and year to year. Once "Operation Striper Egg" is launched, it becomes a 24-hour ordeal.

"I've worked all night on a shocker boat and then had a cold shower and headed for the hatchery to work all day. Then if you're lucky, you miss a night on the shocker boat and can catch up a little bit on your sleep," said Jack Harper, director of the Durant, Oklahoma fish hatchery, which has one of the finest striper projects in the nation outside of South Carolina.

During the six weeks of the spawning season, one fisheries biologist is on duty with each eight-hour shift on the around-the-clock pursuit of the stripers. All additional personnel, including all members of the Law Enforcement Section of the Department are called in to help.

When the stripers are shocked into the nets of the game

and fish department staff and then brought to the hatchery, the hatchery staff first gets the exact time the hormone injection was made. Since 36 hours is the normal time before the females will start to ripen, that enables them to timetable the work in front of them.

The "expectant mamas" are placed in a holding vat six-by-six-by-four feet in size, and the future papas are placed in a similar size tank next to them. Then the biologist, using a glass tube, takes a sample of the eggs from the female striper.

This has proven to be most helpful. It actually is the key to solving the ovulating problem in the striper program. All that the development of this practical technique required was three years of work. unlimited patience, numerous experiments, several guesses, considerable cussing, occasional criticism, and a smattering of luck.

But with the egg sample, the biologist could estimate the delivery time of his mama striper down to almost the exact moment. But even as gynecologists have discovered in dealing with expectant human females, some will and some won't be on schedule.

"We may be right on target with the first two, three or more females, then we hit one that defies normal procedures. When this female is ready, she will deposit her eggs, and not one second earlier," Harper admitted.

"When the female striper is ripe, we have one man holding her head, one her tail and a third rubbing her belly," said Harper. "The rubbing is what does it."

The eggs are deposited in a small plastic dishpan. When the final egg has been laid, the female is placed in another tank of water and the crew picks up a male. The same procedure is used to milk the sperm out of the male over the eggs.

Once the sperm is deposited, water is added. Then gently, very gently, the water, eggs and sperm are mixed so that all of the eggs are fertilized.

The eggs then are placed in a hatching jar about 18 inches in height with a tube that runs down the center. This prevents bubbles rising and carrying the eggs out of the jar.

Now there are striper eggs and striper eggs. The eggs obtained from South Carolina stripers are heavy. Those from California are heavy. The eggs taken from South Carolina stripers are heavy. But the Hudson River eggs have more of a buoyancy. And if the striper eggs come

The Santee-Cooper striper hatchery at full operation.

from Chesapeake Bay, they are so buoyant that you almost have to put a lid on the hatching jar.

While the eggs are hatching, they are counted so that the biologist will have an idea of whether they can expect twins, triplets or a mere million and a half baby stripers. There is a formula for counting that involves centimeters.

It takes the eggs an average of 36 hours to hatch. Once the hatch is completed, the striper fry are placed in an aquarium. And here is where the silk stockings were first brought into use to keep the fry from disappearing down the overflow drain. The same striped bass that will grow into a 60-pound tackle buster, right now is so tiny it is hardly visable.

The 40-gallon aquariums keep a running flow of freshwater, which necessitates the use of some sort of screen to keep from losing the tiny fry. At the outset the fry are fed brine shrimp, a shrimp so tiny as to defy description.

The fry remain eight to 12 days in the aquariums, then they are dumped either into rearing ponds or into available lakes. Both programs are successful. Lakes are utilized where tests have indicated that this type of planting has paid off. But since it is the plan of the hatcheries to stock other lakes, some of the fry are placed in special holding ponds. Here, they are fed for from 45 to 60 days. By then they have reached possibly three inches in size.

During this time they will feed primarily on zoo plankton. After the end of the first two months of their lives, they are large enough to be taken out of the indoor holding tanks and placed in outside ponds.

They are fed first a small granulated trout feed that sinks quickly to the bottom of the pond when spread over the water. When the fry increase in size to five or more inches, a pellet trout feed is used.

Feeding time is a period of frenzied activity. The young fingerlings boil to the top and break water in their anxiety to get their fair share of the food. Oddly enough, this tank feeding isn't programmed into their system. Yet it obviously has little or no effect on the future carnivorous habits of the stripers.

I saw proof enough of this when I witnessed a feeding of the trout chow and then saw one of the five-inch stripers floating on the surface at the rear of the pond, strangled by a shiner then somehow had been left or had gotten into the pond.

Thanks to several decades of experiment, the South Carolina striper hatcheries are capable of handling approximately 100 million babies in a single spawning season. That alone should insure all America of having an adequate stock of stripers planted in their lakes.

South Carolina, with 30 states, Russia and goodness knows who else hammering at her door seeking striper fry, now limits each state to two million fry a year although they each can handle from 10-20 million fry.

When these tiny five-day-old striper fry are received, the states are expected to have facilities for rearing them into healthy fry. This takes from 45 to 60 days. And once they reach that stage, then South Carolina wants 25 percent of the "adult fry" returned.

That doesn't mean that South Carolina, for example, will get a half million fry back out of each two million fry given. That may come to pass some day. But state fishery departments are still in the experimental stages with this gigantic specimen of a fish. Just as the mortality rate was 100 percent in the early work with female stripers, it runs high in the rearing of the fry.

Oklahoma is one of the more advanced states in working with stripers. Yet Jack Harper told me that in the first nine years with almost 70 million fry given to the state, not more than a million were reared to the point of return to South Carolina.

But success is increasing with each passing year. Thanks to the Striped Bass Committee of the Southern Division of

From the aquariums, the 45 to 60-day-old fry go into these tanks until they grow to fingerling size.

the American Fisheries Society, all of the research work of at least 14 states in the division is gathered and then passed around to the member states of Arkansas, Kentucky, Maryland, Oklahoma, North and South Carolina, Georgia, Alabama, Louisiana, Virginia, Tennessee, Texas, Florida and Mississippi.

Since the success of the striped bass hatchery in South Carolina has been established, interest has spread rapidly. Every year, each of the 30 states now sharing in the program send representatives from their ranks of fisheries biologists to study the techniques employed at the striped bass hatchery in South Carolina. And the interest is spilling over to the other states as well.

It was this growing interest that forced South Carolina

The fry are fed the tiny granulated trout feed at 12 days, then they graduate to the pellet size food.

to set up its priority system and limit the maximum number of fry to any given state to two million. And then only if they agreed to return 25 percent to South Carolina. Thus, those states with hatchery setups capable of processing the fry got first priority.

South Carolina bases its priorities for sharing striper fry with other states on three levels. At the top of this list are the states that can rear the 15-day-old fry and return a percentage of the adult fry to South Carolina.

Second on the list of priority are the states that can return another game fish that South Carolina is seeking. South Carolina will accept walleye fry, blue catfish and flathead cats on an exchange basis.

Lowest on the list to receive striper fry are those states

that have nothing to exchange. These states get their fry on the basis of what it costs South Carolina to rear them.

And for those states at least, there must be a great deal of satisfaction in the fact that research has come so far so fast. If South Carolina biologists hadn't been able to improve on that silk stocking drain, being last on this hatchery totem pole for striped bass might have made the stripers just another fish to dream about.

7

Put and take stripers

About 500 miles north of the birthplace of the thirty-third President of the United States is a mile and a half of spectacular scenic beauty. The icy cold, crystal clear, spring-fed waters of the Blue River tumble over the bronze colored native granite in the hills of the Kiamichi Mountains. This granite is the base rock of Oklahoma. Archeologists have placed its age at one billion, three-hundred million years.

Every fall the state of Oklahoma stocks this stream with hatchery-reared rainbow trout. From October 1 through March 30, fishermen who have an Oklahoma fishing license, can obtain a trout permit for an additional dollar and catch a daily limit of six and a possession limit of 12 rainbows.

Normally, these trout are from eight to ten inches in length when placed in the cold, clear waters of the Blue River. There are weekly plantings, but not always on the same day or the same hour. And on at least two occasions during the season, lunker-sized trout in the two-to-four-pound class are deposited as a bonus to the fishermen.

This has been a highly successful program. It is paying its own way. Not only Oklahoma anglers, but out-of-state fishermen flock to the area just north of Tishomingo.

Jack Harper, director of the Durant Fish Hatchery for

the state of Oklahoma is well aware both of the success and of the popularity of the Blue River trout fishing program. It is "put and take" fishing at its finest. Or is it?

What about put and take striped bass fishing? Can an eight-inch rainbow compare to a six-pound striped bass? Can a four-pound trout equal a 30-pound striper?

To each his own. So far, the one fish fly fishermen haven't taken out after is striped bass. Not that there aren't fly fishermen right now making me eat these words.

But put and take striped bass fishing is here. And it will be increased steadily in every state that is planting stripers as time goes by. You see, the striped bass are to a certain extent anadromous. That is they leave salt or brackish water and go upstream for spawning.

But they also can survive quite well in that same freshwater they used for spawning. Maybe they won't reproduce if they don't have just the right conditions, but there is nothing in their nature that won't allow them to be planted in lakes with proper food forage, and then be harvested by the fishermen.

And that is what is going on in a steadily increasing number of lakes across the United States. It is plain unadulterated put and take fishing.

As an example, let us take Oklahoma. The Sooner State has been working with a striper program since 1966. The program started in Keystone Lake, where biologists felt conditions were ripe for reproduction. As it turned out it was a wise choice. Some of the stripers were sucked through the tubes and into the Arkansas River below the Keystone dam. This was a river made navigable through a series of locks. The locks created the necessary flow of water for the stripers to spawn and there have been natural spawns both above and below Keystone.

But in Foss, Grand, Canton, Salt Plains and Lake Murray, the situation was different. Stripers were planted, but there was no natural spawn. But stripers were taken in all of those lakes except Murray. There, only a single planting of adult stripers was made in 1966. Since that time the emphasis has been on the other lakes, plus Lake Texoma. As Oklahoma's striper program grows, Lake Murray will be returned to its program. But it will always be a put and take lake.

Before striped bass can be planted in a lake where put and take fishing is contemplated, a creel survey must be made to determine the shad content. At least three-fourths

Draining a hatchery pond down to net the striper fingerlings.

of the year the stripers depend solely on shad for food. The other three months of the year they dine on may flies or other similar food.

The work of the Striped Bass Committee of the Southern Division of the American Fisheries Society has placed the put and take program in its proper perspective. The reports that come out of this committee are becoming the Bible of the fisheries experts across the land.

For example, a paper on the 1974 study by the group that was presented early in 1975 specifically stated that "some species of clupeoid fish – which is the shad species – must be in the reservoir as a prerequisite for stocking."

The report also said that "lakes which have a salinity

The striped bass fingerlings first go in a waiting tub.

greater than 30 parts per million have been acres of successful introduction of stripers."

What kind of lakes do you have in your backyard? Is pollution becoming a problem? Do sports fishermen feel they are overrun with commercial netters? Are they not too big and not too deep? Is the water temperature on the warmish side?

If you care to believe the Striped Bass Committee, none of those factors have any significance as to whether or not stripers would survive.

It is costly to feed a few million fish in a hatchery. So the put and take method followed in the trout fishing, can't be followed in striped bass fishing. As a result, when

Jack Harper, Durant hatchery superintendent, right, checks finger-lings loaded into waiting truck.

the fish are deposited in a lake, they generally are 45 to 60-day wonders, as the old Army used to call its second lieutenants fresh out of Officer Training School.

What happens to them? Would you believe that by the time the next winter's snow has melted and spring is once again upon us, those same three-inch stripers now are husky two to three-pounders? And a year later they are in the six-pound class.

The average calculated growth in total length at the end of each of the first five years finds the striper stretching from 11.1 inches at the end of its first year to 18.7 inches in the second year; 23.1 inches in the third year; 26.8

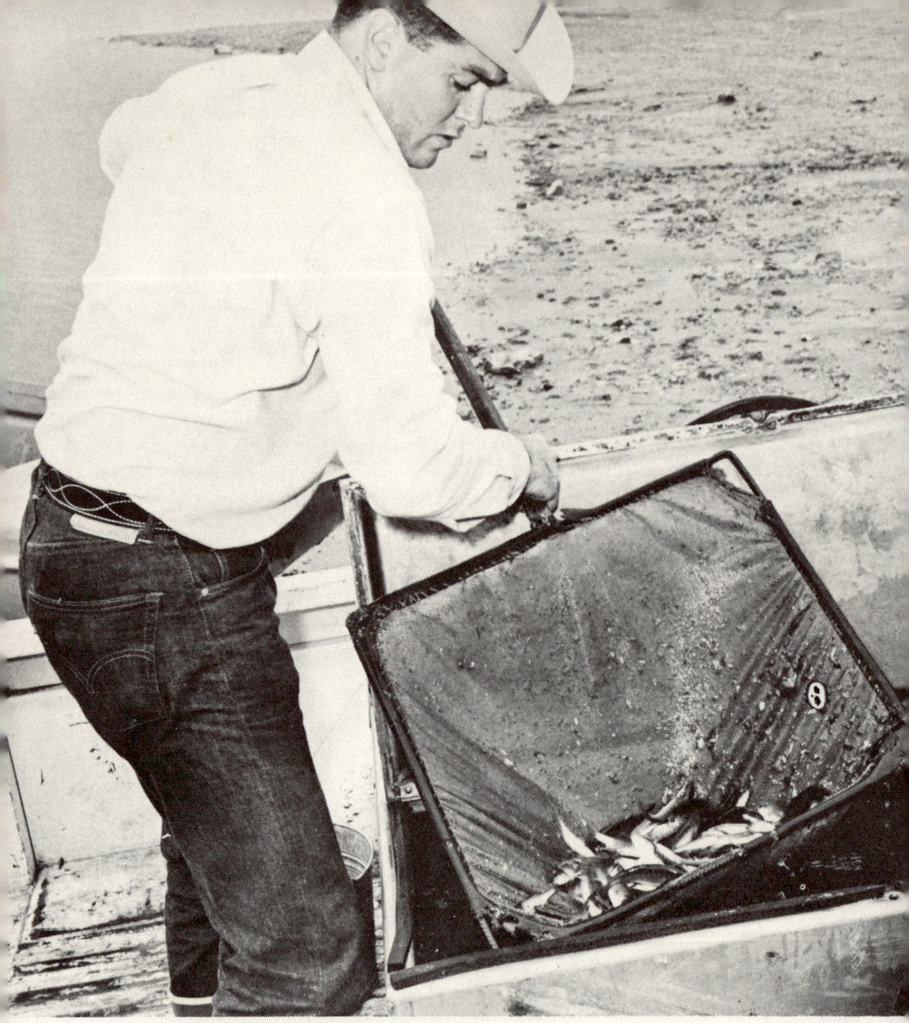

More stripers are taken with this long-handled net.

inches in the fourth year and 29.4 inches after five years.

In pounds, the striper would be in the 12-pound class its fifth year, and its actual size would relate to the availability of food. This is a rather unusual observation, but the fry that somehow went through the tubes from lakes to downstream rivers have a much faster rate of growth. This is because they tend to hover near the base of the dam and when water is discharged through the tubes, the addled shad that come through make dining so easy that the striped bass just have a tremendously fast rate of growth.

Incidentally, the two to three-pound rate of growth per year uncovered in the committee's study is much higher

At the lake the tubs of striper fingerlings are loaded into flatbottom boats.

than the rate reported in the natural habitat of the stripers along the North Carolina coast as studied by Frank Yelverton in 1971. But the committee had an answer to that. In the put and take stocking programs the stocking rates are generally low enough to allow the forage population to remain at a higher level than required by the striped bass for food.

The effects of established striped bass populations upon the forage fishes has been varied. In most cases it has been insignificant or unnoticed. Of all the reservoirs which have an established fishery, only two in the 18 southern states reported drastic reduction in the shad population. But again, these are small 100 and 1,000 acre reservoirs which were stocked at the rate of 60 to 72 fingerlings per acre.

I had a fisherman call me one day and in a mildly complaining manner, tell me that on his last two or three trips to a lake that had been stocked with stripers, he had only so-so luck with black bass.

"I figure those 30-pound stripers are just eating up all of our blacks," he said seriously. "I guess I wouldn't be complaining if I could catch one of those big fish."

That was a natural observation. But it has no merit. In 30 states where the striped bass have been introduced, the striped bass have had no noticeable effect on other native fishes in any reservoir. If there had been any change, especially of a significant nature, the difference would have been discovered during the routine population monitoring by biologists.

The putting of striped bass in inland lakes and the resulting taking of those fish by anglers has had a tremendous economic bearing on the area. Fishermen will drive hundreds of miles for a crack at a big fish. Proof of this is in the fact that 90 percent of all coastal party boats are filled with people from 300 to 1,000 or more miles from the coast.

Looking at that, just think what is happening to inland lakes stocked with one of the prime saltwater fishes, the striped bass. The possibility of landing a 20-pound or more trophy has heightened the enthusiasm for fishing. The fact that shad populations have not been reduced may indicate that the population is being held stable rather than increasing to undesirable proportions. And there has been absolutely no adverse effects on the native game fishes of any of the lakes stocked.

Most of the biologists with whom I visited were firm in their belief that the striper could be the best thing that ever happened to fishing in this country. Since the females produce such a prodigious number of eggs, obtaining the fry is not a problem. And since it has been proven that they can survive and grow even if they are not able to produce a natural spawn, in freshwater lakes, the end result is obvious.

Harper predicted that in Oklahoma in time every lake with sufficient food fish and sufficient size to handle the striped bass, will be stocked. And Harper's prediction was echoed by others.

This planting and harvesting of striped bass has created

a need for a new kind of tackle, from lures to rods and reels. Boats loaded with the latest electronic equipment are the real striped bass masters. All of this cost money, but breathes there a fisherman with soul so dead that he wouldn't part with a little of the long green to enhance his chances of putting that trophy fish in his boat?

By the year 2000, striper fishing should have completely revolutionized the inland fishing industry. Remember all of the excitement generated when the coho salmon put in its appearance? Fishermen from all over the United States flocked to the areas where the coho were running. Overnight, national sporting magazines featured coho lures, coho tackle, coho this, coho that.

Unlike the coho epidemic, striper fishing is here to stay. The coho fever abated. Striper fishing fever will only grow. I was amused at my visit to "Striperland, U.S.A." And what else can you call the Santee-Cooper region? There, every sporting goods store from Charleston to Santee featured nothing but striper lures and striper equipment.

"Don't you ever sell any tackle for the black bass fisherman?" I asked one grizzled old tackle shop operator. "Aren't there any bream fishermen left in South Carolina?"

"They sure is, sonny boy," the operator told me. "But most of 'em had that kind of tackle in the family before these here stripers showed up. They hain't used them much since then, so I 'spect that tackle is still 'bout as good as new."

I got a little different answer in the sporting goods department of a chain store.

"Sure we have it," the young, mod-looking clerk told me with a smile. "But we sort of consider it pornographic tackle around here. You know the kind I mean," he said with a wink.

"Surely you've been in these stores that carry the girlie magazines but you have to ask the clerk where they are?" he smiled. "It is the same with us and any kind of tackle except striper tackle. We keep it under the counter.

"Why it would be unpatriotic to try and push anything but striper tackle in South Carolina, the striper capital of the world. If South Carolina can put the stripers in our lakes to catch, we surely can put the kind of tackle on our shelves for our anglers to catch them with.

"It is the old put and take game," he said.

I thought about that at several intervals during my

Carolina sojurn. And I've thought about it a lot since. I always had a mental image of put and take fishing being much like the Blue River trout, or the catfish farms where you can go and fish as long as your pocketbook can stand the gaff at some 70 cents a pound for all the cats you land.

Then I've been to the trout streams in Colorado where at intervals along the highway you see the "fish farm" signs and you go and fish in a locked-in pond not unlike the fish hatchery ponds.

Why I've even been to sports and vacation shows where there were trout ponds filled with multicolored rainbows and floating chunks of ice to keep the water temperature down. These overfed fish just had to be snagged to be caught, yet frenzied Americans, young and old alike, shelled out their dollars for 15 minutes fishing time.

Put and take fishing.

They all are just that.

But now I have a different image. I have been to a lake where striped bass have been put. And I have fished for hours. I have had successful trips, and I have been blanked. But I didn't come away with that skunking feeling like I did when I paid my dollar and another dollar for my daughter and we couldn't get those tame trout to touch anything we offered. Then I felt unclean. Jilted. You name it, because I felt it.

It won't be that way ever again. It won't be for me, and it shouldn't be for any angler anywhere. Because with striped bass planted in 30 states in 1975, there isn't anyone anywhere who is too far away not to have a chance to try his luck at some put and take striper fishing.

8

Con Ed caper

When historians look back on the mid-1970s, the economists are going to possibly draw some rather dastardly deductions about the operation of Consolidated Edison. Whether they are looking at the rising costs that users of electric power manufactured by Con Ed were charged or at some of the editorial cartoons that lampooned the utility giant, the picture is going to be strikingly similar.

But the ecologists are going to take a look at what Con Ed was doing and they are going to draw some completely different concepts. You see in that great inflationary era of the $oaring $eventies, when you had the unique economic situation of a runaway inflation coupled with a rising rate of unemployment, any and all utility or oil or sugar companies were prime targets.

People had become completely attached to being kept cool in the summer and warm in the winter. And all of that took a great deal of power. Of energy. It took utility giants like Con Ed to supply the energy. And to obtain the energy sometimes it took a lot of natural resources like oil and coal and water.

And it was the extremely heavy use of water that got Con Ed into more trouble with the conservationists and the ecology people than with their customers for their high prices for power.

Con Ed's hydroelectric operation on the Hudson River was taking out tremendous volumes of water. When this action occurred during the spawning season of the striped bass that used the Hudson River for their freshwater base, the ecologists came unglued.

Con Ed ran studies. They determined that there was some loss in the striper population. But they also arrived at the conclusion that before there was a loss, either the spawning female striper on its way up the river or the eggs and larvae on their way down, had to face impingement against the power tube screens to suffer damage. If the stripers were sucked up into the tubes, then the loss was total.

Early in 1972, Dallas-based Texas Instruments, a world-wide electronics manufacturer, opened a rather unique branch. It was labeled TI's Ecological Service. Staffed with the finest biologists and ecologists available, the purpose of this service was to take ecology problems that confronted industrial giants and to try and find a solution.

Con Ed had been toying with the ideas of building a fish hatchery along the banks of the Hudson. So it contacted TI and turned its ecology problems over to them to see what TI could find.

When TI ecologist Phil Welsch and his crew arrived, Con Ed was at a low point in its public relations with its customers. The users of power were caught in the economic squeeze of higher prices. The nuclear plant was destroying fish in the Hudson and an ecology group was baying at Con Ed's heels.

But immediate publicity on the hiring of the TI group helped. Welsch and his staff set out gill nets along a 40-mile stretch of the Hudson that they determined was the "spawning run" of the stripers coming out of Chesapeake Bay.

They caught the brood stripers in front of Con Ed and as far as 20 miles in either direction.

The brood stripers weighed in at from 18 to 40 pounds. And they produced a total of 15½ million fry.

When caught, they were given an immediate hormone injection to induce spawning then transported by air to a federal fish hatchery in North Carolina. As the hormone takes from 30-36 hours to work, there was plenty of time to transport the fish to Carolina before the females were ready to deliver their eggs.

Even so that is a ticklish job. It was most difficult to

Texas Instruments' ecologist Phil Welsch helps load a waiting plane with Hudson River stripers.

keep the brood stripers alive long enough to strip the eggs. The eggs that were taken were hatched into fry and most were kept at the federal hatchery until they reached fingerling size.

However, about 4½ million of the three-day-old fry were flown to the Oklahoma State Hatchery in Durant, where they were raised up to fingerling size. This was a backup precaution taken by Welsch just in case something happened to the fry he was trying to rear in North Carolina.

Transporting the fry by air required a lot of work. Welsch found that the best way to transplant the striper fingerlings was to place them in double plastic bags at the rate of a quarter pound of striper per gallon of water. The plastic bags were placed inside styrofoam cartons and filled with oxygen. Then the cartons were packed inside cardboard boxes, which at final count contained less than 60 stripers each. With all this care, it took 77 boxes loaded in a chartered plane to take just 4,620 of the fingerlings from Durant back to the Hudson River.

The stripers were taken from the Oklahoma hatchery at 4 o'clock in the afternoon and 11 hours later they were swimming free in the Hudson River near the Con Ed plant. It took some precision timing to have the charter transport plane carrying the cartons arrive at the Dallas-Fort Worth Airport in time to meet an 8:54 p.m. plane for New York. This plane arriving in New York at 12:46 a.m. was met by another charter plane which took the stripers on to the Con Ed plant site near the river.

Welsch described this fingerling stocking as mitigation for the loss of that percentage of stripers that did find their way into the screen and tubes during the Con Ed operation.

"Several things happen during spawning season to the spawning stripers that have come up through the Chesapeake Bay and on up the Hudson River," said Welsch. "For some, they are banged against the screens by the current and wind up sucked through the tubes. Even the striped bass larva could undergo the same treatment."

The study by the TI ecologists uncovered some rather interesting observations on the striped bass. Welsch is convinced that from his research there are different "races" of striped bass, possibly a northern striper, an eastern striper, an Atlantic striper, and a Pacific striper.

And because some stripers, such as the Hudson River strain, don't make a spawning run from the sea, to find brackish and later freshwater, finally returning to their saltwater haven, Welsch agrees there must be "races" of landlocked stripers, too.

By the mid-1970s, Welsch's Texas Instrument ecologists had a lot of projects underway with the Power Authority of New York and of Kingston, New York and with Detroit Edison and Niagara and Northfield, Connecticut. Naturally, not all of these projects dealt with stripers, but the list does give a pretty good idea of the scope and expanse of work that this new division of TI was doing to improve the lot of the outdoorsman.

With the kind of protection Welsch developed for transplanting stripers he found they could be hauled by charter

The striper fry were taken from the Hudson River, reared at a Durant, Oklahoma, hatchery and now are being flown back to the Hudson River.

aircraft direct from the federal hatchery in North Carolina, from the Oklahoma fish hatchery or from a California hatchery with a minimum loss. The usual procedure was to fly them by charter plane to the nearest airport where they could be transferred to regular airline service for the flight to New York.

The price Con Ed pays for this program is not only in hiring the TI ecologists, but in giving a percentage of each fingerling harvest to the hatcheries that do the work. This probably is a cheap price to pay as compared to setting up their own operation.

But this has to be an interesting sidelight on one of the real utility giants of our time.

So successful was Welsch's project that Con Ed was expected to have its own hatchery in operation along the banks of the Hudson River before the end of the 1970s. In this manner, Con Ed then can take out the proper number of brood stripers during spawning season, hatch them right on the banks of the Hudson, and then return the first 15 to 150 million as fry, and follow it with fingerling planting.

Welsch said the reason a percentage were returned as

fingerlings was to give assurance that the fish put back in the Hudson survived. Each fingerling that was placed back in the Hudson, and 102,000 were in this particular year, had a stainless steel wire .001 of an inch long in five assorted colors placed in its lip.

Then each of the stripers had a fin clip attached. Recovery ranged between three to five percent, but since the data on each fish gave its size, date of release, area in which it was released, etc., the TI group was able to ascertain just what had happened to the fish returned into the Hudson River.

While the striper project is what brought the TI Ecology Service to the attention of Con Ed, Welsch said in the projects his crew had going in the north and eastern part of the United States, 99 different species of fish were studied.

"We checked everything from their size to their sex habits in our research programming," he said. "All was aimed at improving fishing in rivers, lakes and streams that may or may not have been damaged by industrial use of the water."

By the time TI finished with Con Ed, there was a definite shift in the image of that Eastern utility giant. No longer was it the ogre of the past. Instead, it had every chance of taking its place as proudly on the ecological front as it had done in scientifical circles when it became a forerunner in utilizing nuclear power to operate its hydroelectric plants.

9

Hybrid stripers

What do you get when you cross a striper with a sand bass?

For one thing, you get a hybrid bass that has irregular lines somewhat patterned after the sand bass, but as dark and distinct as the striped bass. And you get a fish that many anglers feel, pound for pound will outfight the striped bass, although perhaps not quite the table fish of the striper.

The most successful chapter in the hybrid work has been written by the state of Tennessee. There, the game and fish commission crossed the striped bass female with the white bass male and produced a fish that grows quickly – and to a respectful size – and is adaptable to lakes enjoyed by either parent. In the mid-1970s the state record for stripers was 33 pounds against an 18-pound state record for the hybrid.

Tennessee made its initial crossing of the striped bass and white bass in April 1965. The most successful cross has been with the striper female and the white bass male. The reverse cross has a very poor survival rate.

Today, biologists are convinced that many lakes in the nation that have large populations of white bass and which have added striped bass are getting a natural hybrid spawn. But whether they do or not, it is no secret that the hybrids, unlike both parent species, handle well under hatchery conditions. They have been held in concrete raceways for as long as seven months successfully, while with stripers it is often difficult to keep a female brood striper alive long enough to strip the eggs.

Tennessee became interested in the hybrid after repeated failures to establish a fishery with the striped bass. While they couldn't rear striped bass, maybe they could come up with a fish similar to the striped bass but more adaptable to inland reservoirs.

The idea of the cross was not new. It had been discussed by marine biologists for several years. In fact, in 1963 biologists at North Carolina's Weldon hatchery announced they were planning to cross the male striped bass with the female white bass. For reasons unknown, this experiment was not conducted.

A year later the technique of injecting striped bass with hormones to produce viable eggs while male white bass were still ripe was developed. Since it made hybridizing an easy process, the experiments were initiated by the Volunteer State biologists in 1965.

But they did little to indicate the future of this process. Fry were produced on the first try, but efforts to rear them in an aquarium fizzled. The second attempt came 10 days later. Some 155,000 fry were hatched but a dieoff three days later cut the number to 45,000. Three thousand were lost when exposed to water having a high pH. Ten thousand were stocked in two farm ponds but there were no survivors. The remainder were put in an aquarium but they also died.

The third hatch nine days later produced 80,000 fry. They were held in an aquarium for 48 hours without a significant loss. Half of them then were put in plastic bags and transported from Moncks Corner, South Carolina to Tennessee. Losses on the trip were less than one percent. Once in Tennessee they were placed in a fiberglass tank, where on the eighth day of their lives they were observed feeding on zooplankton and appeared to be in excellent condition. But the next day they died. The blame was placed on the 80° water temperature.

The other fry had been released in the Frog Pond. They are the only known survivors of the three crosses. The Frog Pond is an isolated area of Cherokee Lake, a 30,000-acre reservoir created as a flood control and hydroelectric project. It has an annual fluctuation of from 70 to

90 vertical feet. The lake is in eastern Tennessee where large recessions have been formed in the limestone substrate. Thus, potholes are formed as the water level of the Cherokee rises and falls.

The Frog Pond is such a 500-acre pothole. When the water rises, the fish migrate into the Frog Pond through two large crevices. Since it generally is underpopulated when this occurs, the survival rate of fry is usually high. Six months after the hybrid fry were put in the Frog Pond, recoveries were made and a total of 5,017 hybrids up to 8.4 inches were taken.

Four months later nine hybrids weighing up to one pound were recovered from a pothole several miles from Frog Pond, bearing out suspicions that some of the hybrids had escaped during the high water.

With that success, Tennessee duplicated its Moncks Corner project a year later, only this time the hybrid fry were placed in specially prepared outdoor raceways. Losses here were less than one percent and this is the first major stocking that Tennessee made into Cherokee and Norris Lakes.

As the project was pushed, the state tried the reverse cross and then tried back-crossing using hybrid males and striped bass females. But the results were 50 percent badly deformed fry.

The most often asked question of today's fisherman is how do you tell the difference between a white bass, a striped bass and a hybrid? At first glance, the hybrid is a blend of its two parents. The nape of the hybrid is more prominently humped than striped bass, but not as much as the white bass.

Coloration is quite similar. The hybrid is olive-green along the dorsal region, becoming lighter on the sides and white ventrally. There are usually four stripes above the lateral line and four below. The stripes are almost black, and unlike the white bass, they are wide and are prominent to the posterior end of the peduncle.

The arrangement of the patches of teeth on the tongue of the hybrid are identical to those of the striped bass. There are two long patches near the anterior end and two small parallel patches near the base of the tongue. The white bass has the long patches near the anterior end, but the patch near the base is usually a single round spot. However, white bass in Cherokee Lake weighing more than 2½ pounds frequently have a divided patch on the base of the tongue. You might check your own white bass the next time you catch one that size.

The general appearance is sufficient to distinguish the adult hybrid from the striped bass and the white bass. However, with reciprocal crosses and back crosses entering the picture, it will become more difficult. And in lakes where hybrids have been planted, these other crosses can come from a natural spawn.

But, why go for a hybrid that you know will get to 18 pounds, when you can shoot for striped bass that have topped 65 pounds in freshwater?

Remember those 35,000 hybrid fry that were planted in the Frog Pond? At the same time, 3.5 million striped bass fry were released in the main body of Cherokee Lake. Yet the confirmed catch by sport fishermen has been more than 80 hybrids for every striped bass! That means that the survival of the fish was for the positive benefit of the fishermen and that is what it is all about.

Survival rate of adult striped bass and white bass isn't very high when they are placed in concrete raceways for any length of time. Adult hybrids have been held in the same kind of raceways for up to seven months without losses. And during that time by being fed a diet of live fish, they increased in weight by almost 50 percent.

Speaking of growth, what about those stripers and hybrids that were planted in Lake Cherokee? The plus for the hybrids came in the first 18 months. They grew to about four pounds, as compared to slightly under four pounds for stripers. And at 27 months the biggest hybrid was 5.2 pounds to 4.4 pounds for the striper. But then the striper moved rapidly ahead.

The hybrids offer the best sport for anglers starting in late February or early March, during a time when in Tennessee lakes, at least, there isn't much other fishing activity. As a result, this makes the hybrid especially attractive.

At the age of three years, the hybrid has grown to three times the size of the white bass of the same age, and they are powerful fighters. Fishermen regard the hybrid as an excellent game fish and rate its food quality at under the striper but above the white bass.

The hybrids feed on fathead minnows, goldfish, threadfin and gizzard shad, carp (and that's a plus for every lake), bluegills, mosquito fish and golden shiners.

Today, biologists are convinced that there are lakes and areas where the hybrid is by far more suitable than striped bass. The cross between the white bass male and the striped bass female appears to be one of the most successful ever produced. What Tennessee has done has been widely circulated in biological fields and today, many states with striper and white bass programs are duplicating the efforts.

And of course, there is always Mother Nature around with her own way of producing natural hybrids. Texas biologists are confident that this is happening in their state and they are more than delighted with the results. Pat Mayse Lake near Paris has been heavily stocked with hybrids, while other hybrid stockings have been made at Lakes Nasworthy, Garza-Little Elm, Bastrop, Jacksonville and Martin Creek.

One of the reasons for the success of the hybrid program is the fact that it has such a rapid rate of growth. With some hybrids exceeding five pounds – the record size for the white bass – in just a couple of years, that gives the hybrid a significant advantage over the white bass as a sport fish. The white bass feeds predominately on threadfin shad, while the larger hybrid will help keep down the gizzard shad population as well as carp.

When anglers start running up against 10 and 15-pound hybrids, the selling job of the biologists is over. Fishermen are impressed. Not only by the size, but by the fighting ability. And when the angler carries that fish home and learns that it is tastier than the white bass, he's 100 percent sold. Of course, it will be nice if the hybrid does the shad control work that it is cut out for, too.

Since efforts to back-cross the hybrid has resulted in mostly deformed fish, this could create a whole new problem for lakes were hybrids are stocked. Deformed fish such as those produced by back-crossing the hybrid male with the striped bass female would not be desirable sport fish.

One of the most pleasing advantages to the hybrid as far as biologists are concerned is their fast growth rate during their first two years. This growth starts in a hurry. By the time striped bass reach two to three inches, hybrids are five to six inches. They start taking feed earlier, are easier to rear and have a significantly lower mortality rate.

The back-crossing introduced some weird results. Some

were sucker-mouthed to such a degree it is doubtful they could effectively prey on other fishes. There were also pug-nosed hybrids, and others with curved spines that made them banana-shaped. And while Tennessee was wondering if this was happening only to them, they got word that Texas' hybrid back-crossing was faring no better.

As time goes on, the hybrid will become somewhat less important in larger lakes and in states where stripers can be successfully planted. Now that the techniques for developing and rearing stripers has reached an almost foolproof point, the nation should have numerous inland hatcheries by the end of the 1970s. But at the same time, there always will be a place for the hybrid. It has proven to be a desirable fish. It has a high survival rate. A rapid growth. A fighting quality. Excellent food value.

And in a time when there seems to be no end to rising costs, hybrid fisheries can be created with less expense and effort than required for striped bass.

But surely we can come up with a better name for the hybrid than hybrid. How about White Striper or Speckled Bass?

10

Longhorn stripers

The best striped bass fishing in Texas is in Oklahoma. Let me put it another way. The best striper fishing is in Lake Texoma.

And until the Colorado, Brazos and perhaps the Trinity River Systems produce a natural spawn, that situation won't change – even if then.

Yet, Texas hasn't planted a single striped bass fry or fingerling in its best striper bass lake.

The reason is simply that the Sooner State, which has waters in 70 percent of Lake Texoma, took the lead in striper planting in the big border lake 75 miles north of Dallas. This was way back in 1966. Since that time, Oklahoma has planted more than 1½ million stripers, from fry to adults.

But then that is hardly an issue since by the mid-1970s at least two natural spawns have been confirmed by the Oklahoma biologists. A 10-pound female can produce a million eggs, and by the time you get to a 40-pound female striper you are talking about 4½ million eggs. So you can see that with a natural spawn, the striper population of a lake can reach millions or billions in no time.

Naturally, not all of the eggs hatch nor do all of the fry survive, but the survival rate is high enough to create some magnificent fishing conditions. In Texoma, it is felt that most of the stripers go up the Red River to spawn. Since the south bank of the Red is the state line, that means that all of the hatch from a normal spawning period produces only native Oklahoma stripers. And if some of the brood females make their way up the Washita River for spawning, all of that river is in Oklahoma so that all of the stripers produced from a natural spawn are Okies.

Stripers have been released in 13 Texas lakes. They include Lakes Nasworthy, Garza-Little Elm, Pat Mayse, Bastrop, Jacksonville, Martin Creek, Spence, Granbury, Whitney, Travis, Canyon, Toledo Bend and Amistad.

The best survival rate on stripers has been in Lakes Spence, Whitney, Granbury and Mayse, according to Texas Park and Wildlife officials. And Texas biologists concentrated their planting of stripers in Lakes Granbury, Whitney and Travis during the mid-1970s. The reason was obvious.

If the program continues as successful as it started, these lakes can furnish a source of fish to intensify stocking efforts in other lakes.

Texas utilizes Parks and Wildlife Department hatcheries at San Angelo, Lewisville, San Marcos, Heart of the Hills, and Possum Kingdom, as well as the U.S. Fish Culture Station at Tishomingo, Oklahoma for its striper fry and fingerlings. Thus, even Texas is moving into Oklahoma to get assistance for its striper program, albeit at a federal hatchery.

Despite the failure of some transplants, the program works like a charm in other lakes. And since they have definitely reproduced in many states that have the proper upstream or downstream conditions, Texas expects to be one of the premier striper country states. Both the Colorado and Brazos River systems have the proper quality and quantity of water, the two key factors to a successful spawn. The Trinity River has the water, but the quality is not the best.

Texas got its brood fish from out-of-state sources prior to 1973. But in that year, state biologists were able to obtain brood from Lake E. V. Spence, located north of San Angelo. The subsequent fry and fingerlings raised from the Lake Spence brood stock have been the basis for the Lone Star State stocking programs since the program was instigated in 1973.

Most of the brood stripers taken from Lake Spence weighed in the 11 to 16-pound class, which means that at a maximum they could produce from one to 1½ million eggs per brood fish.

Texas also is working on a continuous cross-breeding program. At the outset, Lakes Bastrop and Pat Mayse were selected for this work. During the mid-1970s, there was more concentration on Lake Pat Mayse. The cross is made by using the striped bass eggs that have been fertilized with white bass milt. This is an interesting offshoot of the striper program. Kentucky Lake in Tennessee has produced hybrids up to 18 pounds. Since the record white bass was five pounds, two ounces at the time, you can see that this is bringing another game fish to a lake.

Many biologists feel that the hybrid bass is the answer for putting a big fish in a little pond. While stripers quickly grow into 20 or 40-pounders, hybrids remain well under that although double or better the size of the smaller fish involved in the hybrid program. Hybrids therefore make a perfect fish for stocking in smaller lakes that doesn't have the shad food population necessary to maintain the larger stripers. Bastrop and Pat Mayse Lakes may pretty well fit that bill.

The two most popular methods of fishing for stripers is either trolling or jigging. Eventually, when the striper population builds up, you can expect surface fishing when they churn the water much like the sandies during feeding periods. This occurs regularly in Santee-Cooper, but it took a decade or more after the fishing started for this to happen.

Lake Texoma undoubtedly was the first lake to offer surfacing stripers in Texas. During the early 1970s following the initial planting in 1966, an occasional striper started appearing. But by the mid-1970s, striper fishing was rapidly becoming the predominate sport of anglers. And if the Santee-Cooper fishing cycle is followed, before the '70s runs its course, the stripers should be tearing up the surface during feeding periods in the warmer months of the year.

Freshwater striper fishing is the only kind of fishing I know that will detract a black bass purist. A black bass angler has to tie into no more than a couple of the huge stripers before he becomes addicted. Striped bass hit extremely hard, much harder than any freshwater fish I've ever encountered.

I was trolling Texoma one day for stripers and we had three aboard, all in the 10-pound class. We had hit a striper alley and our fish locator showed that it was loaded. At that time the limit was one. There were two others in the boat so I suggested that we give one away and try to land another. We pulled up next to a boat with three men jigging.

"Any luck?" I asked.

"We gotta few," one of the men replied.

"We have a limit and wondered if you'd like a fish?" I continued.

"We're doing all right," the man said, but as I reached over in the live box and pulled out the striper, his eyes got as big as saucers and he stammered "but gawdamighty I'll sure as Hell take one like that!"

The trio admitted they knew about the striper in Texoma, but at that time had no idea how to catch them or what with.

While this is covered in detail in another chapter, I'll just point out that stripers are caught on a variety of artificial lures including both deep and medium runners. Jigging or slabbing is another favorite way. If you use a jig, you probably will score better by adding a streamer fly or a white plastic grub worm.

The free spool reels and good stout black bass rods or your open face reels that you double with when you are doing some coastal fishing, generally will suffice.

Now that Texas has a verified survival of striped bass in Spence, Granbury, Whitney, Travis, Canyon, Mayse, Garza, Bastrop and Nasworthy, this doesn't mean that the fish have produced a natural spawn. It simply means that the striped bass have adapted to these freshwater lakes, that there is ample food in these lakes to provide for the stripers, and that they are alive and well and growing.

One of the nicest things about the stripers is that there is no need to provide special cover for them. Unlike their saltwater cousins, the red snapper, they don't require special housing. Old car bodies have provided snapper banks on the Texas Gulf Coast that has proven a bonanza for snapper fishing. But the striped bass are quite happy with a slick lake bottom.

The stripers antics are more akin to the white bass, and no one is more familiar with the habits of the white or sand bass than Texas anglers. The stripers are a school fish and tend to herd their food fish or baitfish in true Texas cowboy fashion. They generally feed into lee sides of the shad they have rounded up.

There is nothing that any of the researchers on stripers have found that gives any indication that the stripers will feed on any other game fish. They select shad, threadfin and gizzard, for from nine to 10 months out of the year. Then if you have a good mayfly crop – and Texas lakes generally abound with this striper delicacy in the early spring – this becomes their change-of-pace diet.

Because of the sensational success Texas had had with white bass, marine biologists are joined by fishermen in rooting for a similar success with the striped bass. And there is every indication that the program will succeed. Because of the prolificy of the stripers, even if every striper put in Texas lakes has to be hatchery reared, the program can succeed.

If Texas fisheries people have had any complaints from black bass fishermen over their fears that striped bass may hurt that kind of fishing, they need do no more than cite some Santee-Cooper records set since the advent of the striped bass in those lakes.

Santee-Cooper has owned the world record for black crappie at an even five pounds. A world record for channel catfish of 58 pounds. A state record for a largemouth black bass of 16 pounds, 2 ounces, (and that far outstripes the Texas state record for black bass at this writing). A state record for pickerel of six pounds even. A state record for the white crappie of five pounds, two ounces, which is just an ounce away from the world's record. And, the world record for land-locked striped bass of 55 pounds.

If the two lakes in the Santee-Cooper reservoirs, Lake Marion and Lake Moultrie, can produce that many record fish of that many different species, it should pretty well answer all doubts of Texas fishermen on what to expect in Texas lakes that have been stocked with stripers.

I am sure as time goes on and Texans become more accustomed to these Texas-sized fish in their own lakes, that they'll also learn new and different ways to land them. At various times of the year when they chase the shad in Santee-Cooper to the surface, they are caught with surface plugs or shallow runners. And the easiest way to spot them is to watch for the gulls, that had rather dine on the shad addled by the stripers, than to fend otherwise for themselves.

But as most Texans who fish realize, the presence of

gulls generally indicates sand bass feeding. But all of that started changing in some lakes in the mid-1970s. Just as anglers have caught black bass feeding with sandies, now they are catching stripers. Nothing too big at this point, mostly one to three or four pounds. But the very fact that they are catching them indicates that in the years ahead, the stripers may be running in their own schools.

Since the stripers are a schooling fish, they come out of deeper water to feed. If you can't catch them on top or at a shallow depth, stick around and go deeper either trolling or jigging.

Because a lake has a lot of trees or underwater stumps in it and has a fine reputation as a black bass lake, doesn't mean that stripers can't be compatible companions to the blacks. Lake Moultrie is a clean bottom lake, while Lake Marion, the other half of the Santee-Cooper tandem of lakes, is filled with underwater brush and stumps in addition to having trees jutting out of the lake. And stripers are caught in like numbers in both of the South Carolina lakes.

One Texas marine biologist apparently has looked a black bass angler squarely in the eye on the matter of whether the stripers might hurt the blacks or vice versa. This biologist has come out for the record to say that "stripers have been known to grown to 50 pounds or larger on a diet of shad and other forage fish, but their feeding habits have little effect on native species such as largemouth black bass."

11

Fishaholic

The cream-colored station wagon pulled up in front of the house about a car's length past the drive. It stopped abruptly, then was backed expertly into the drive, coming to a halt about a foot shy of the overhead garage door.

A blond Swede of medium-height and reddish complexion, in a pin-striped business suit stepped out of the car. He reached across the seat, picked up a brown briefcase and then walked into the house.

Less than 10 minutes later, the overhead garage door opened. The same man, this time wearing a tan jump suit and with a billed cap on his head, picked up the end of a boat trailer and rolled it slowly down a slight incline forward to the trailer hitch on the back of the station wagon. He made the connection in a matter of fact manner, as if it had been something he was accustomed to doing. He plugged in the lights, then walked around and got into the driver's seat.

As he turned the corner, the elevation was such that he could see the huge border lake. The water had a nice chop to it, with white caps flaking at intervals. The driver of the car, Texas Instruments, chemist Len Andersen, smiled.

"The fishing might be pretty good," he thought as he headed his car toward the launching ramp where he would put in. He was no more than four minutes away from his house when he backed his boat into the water.

Len Andersen is a new breed of American. He has been captivated by inland America's new super fish – the striped bass. He can't wait to get home from work to go fishing for the stripers. He spends most of his weekends chasing the striped bass.

Len Andersen doesn't know it yet. But he is a fish-aholic. And it is all because of the striped bass. Before the advent of the stripers, sure Andersen went fishing. He enjoyed the lunker black bass and the hard socking sand bass. In fact, he is as good a fish chef as he is a fisherman. He has a sense of humor, too. He figured out a recipe for pickling the sand bass and he put it down on paper in chemical terms. When guests oh and ah over his pickled sand bass and ask for the recipe, he gives it to them. It is completely accurate, but the metric terms used are rather strange to most cooks. But as he told them, "we're coming to the metric system sooner or later anyway so you might as well start now."

The striper season on this lake is closed during the spawning period. And then when it reopens, the fish have moved about so much they are hard to find. And so there are many water hauls in the early spring in Len's pursuit of stripers. Oh, he makes the best of it. He catches a few lunker bass. And when he runs up on schooling sandies, he'll harvest 50 or more for fillets. It is great sport. He won't knock it.

In fact, a few years earlier he could tell you where all of his favorite black bass holes were. He could tell you how he could spot the sandies. He bragged about his lunker blacks and his huge strings of sandies.

But he doesn't talk too much about them anymore.

He was on the phone about 1830 hours on June 12 of this year. It was the first time he had called since mid-winter.

"I just thought you'd like to know that I have just caught my first striper of the year," Len said. "I took him at 1750 hours on a white Slab. He weighed four pounds."

Len fishes in what he bought as a bass boat. It has three upholstered swivel seats. There is a depth finder that locates the dropoffs that create the striper alley areas that he and all striper fishermen seek out. He has both a large 65 horsepower motor to get him to the striper site on time, plus a trolling motor that is a relic of his black bass days. Oh, don't get me wong, he still will try for a black, but it is the stripers that have caught his eye.

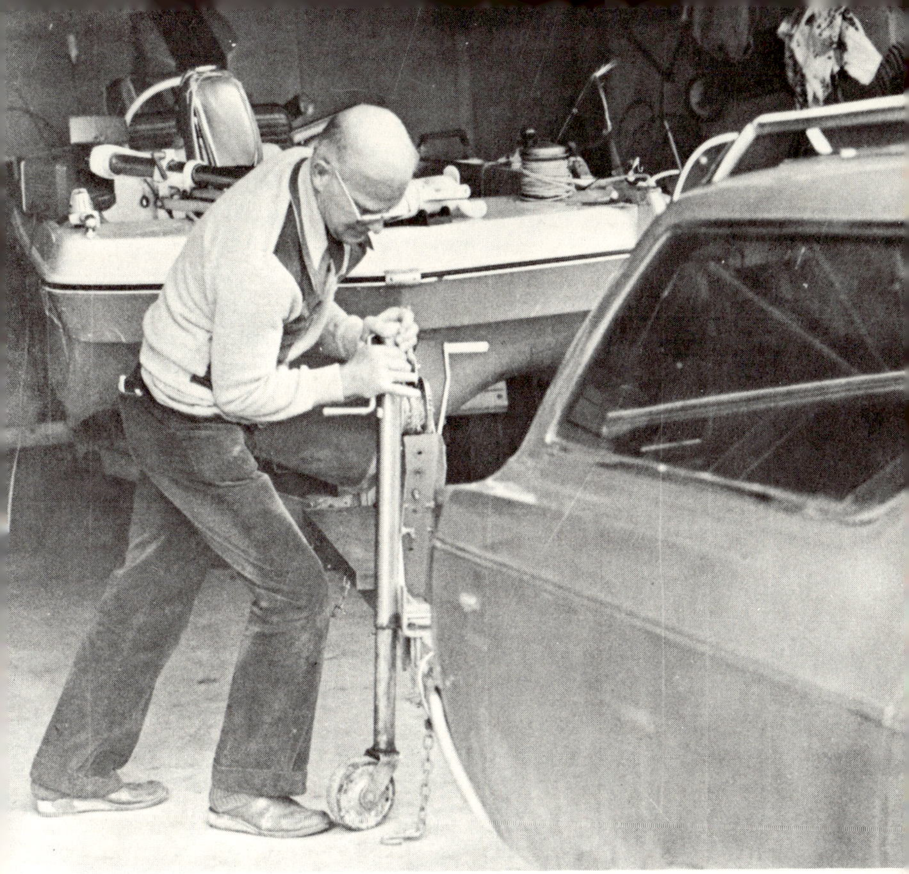

Lennert Andersen rolls his boat down from the garage to hook it on the station wagon. Five minutes after he hits home, he's ready to go out after the stripers.

Len is a tea sipper. His thermos jug is always in his boat and it is filled with a strong, well steeped hot tea.

I remember being out with Len and a third party on one occasion. Len offered tea. When he poured a cup to the invited guest, he took a sip and spewed tea from one end of the boat to the other.

"Hot tea!" the Texan roared. "In this country when you offer a man tea it's got ice in it!"

It was July 12, one month to the day, when I got a second call from Len. This time the striper hit 50 minutes earlier. But it still was just a four-pounder. The bait was still a white Slab.

"They're on the bottom, You have to jig for them," said Len.

"Hey!" the voice was unmistakably Len. The time was

10 days after his last call. But this time there was an urgency of excitement in his voice.

"I caught an eight-pounder this evening," Len said. "I took it on a magnum Hellbender. The one with the white belly and black back."

For Len, that was the official opening of the striper season. What had gone on before didn't count. Now that he had his first striper trolling, he knew he was in for a great year.

"I took him right in the striper alley in front of the dam," he said. "You know, I was a little disappointed in his size. I expected a much bigger striper out of that hole."

In spite of all of the excitement over that eight-pounder, it was three weeks later before Len caught his next fish. It hit at about the same hour on the same lure but weighed only 5½ pounds. But on this next pass down the same striper alley he landed a 12-pounder.

"That is more like it," he dutifully told me. "I had David along so we could keep both of the fish."

David, Len's son, was along on a Saturday morning in July and they were back at the dock at noon with a pair, but they were 3 and 3½-pounders. That afternoon he took out two men from TI. He just ran the boat. He let them fish. When they left they were planning to buy a house by the lake, a striper boat fully equipped, and all of the required gear. The reason? Two stripers, one 9¼ pounds, the other 10½ pounds.

Len felt it was just his due when he went out alone the next day and caught a 10¼-pound striper. He still was using the black back magnum Hellbender, with either a white or pearl belly.

"I almost lost it," he admitted.

"I led this striper right up to the boat," he said. "I was sure that it had been played enough to be tired. But when I raised my rod tip and reached out with the net, it sounded once more.

"It kinda caught me by surprise and before I knew what had happened, I had let it get a little slack. But fortunately, I was able to get the landing net back in the boat and quickly pick up the slack and the fish still was on. It had made a short dive then headed back up. That's when for a few seconds it was out of my control."

Len and David scored on a pair of 4-pounders three days later after weather had kept Len off the lake. He felt the catch was an omen, because he really had little hope of catching a fish at that time with the barometer still falling and the weather still unsettled.

It was a week before he got back out. But on September 1 he took a 9½-pounder. At this point in time he had landed 14 stripers since June 12 and all but two had hit at between 1800 and 1900 hours.

Len celebrated Labor Day weekend with a 10 and a 12-pound striper. That gave him something to talk about in the weeks ahead, since he was sent out of state on temporary duty for three weeks.

On September 29 he was back and right at 1800 hours a 9¼-pounder zinged his black back magnum Hellbender. Two days later he caught an eight-pounder, and after another two-day miss, he scored with a 11¾-pounder.

One thing about the fish Len brings home. It doesn't stay around the house long. You can bet that it will be an entree the next day or that weekend. If the Andersens are having company, Len might smoke several stripers as he catches them. Smoked striper makes a great pickup food for a party when served cold.

There was only a 5½-pounder waiting for Len on October 9, but he won't forget Double Ten day in a hurry. At 1835 hours his reel started singing. He shut the motor and got down to the business at hand.

"I knew it was a big striper from the way it hit. I troll pretty fast you know and when this fish hit, it almost stopped the boat."

It took Len the better part of 30 minutes to wear the big striper down. Had there been a fishing companion along, he could have landed it in half the time.

"But when you have to manage your rod and reel in such a manner as to not give the fish any slack, and still come up with the net, it takes a lot of doing with a big striper," Len said.

This one weighed in at 14 and a fraction pounds, his biggest one of the year to date.

He came right back the next day at 1920 hours to land an eight-pounder, and after missing a day, he scored with a 11½-pounder a day later.

The weather kept him off the lake for a week. But he was back with two companions on a brisk October Saturday. Len was still explaining to his guest fishermen how it all worked when one of them complained he was hung.

The time was 1805 and the "log that came alive" was a 10½-pound striper.

Len played the perfect host and his other guest scored the next time they made a pass down this striper alley. It was only 5½ pounds, but to the angler it was a whale.

At 1930 hours when dusk had settled on the next to last daylight saving time afternoon, Len got his striper. It was a perfect companion to the first catch, weighing in at 10½ pounds.

Len was alone on the next day. It was 1815 hours when his striper sounded. It had hit the green back magnum Hellbender this time. But it was landed as neatly as the rest. He came in at an even 10 pounds.

Darkness covered the lake the next evening, but the sun had made a brilliant exit. The pinks and reds that illuminated the western horizon kept Len still working. At 1840, which would have been 1940 a day earlier and the latest hour he had gotten a strike all year, he had his fish on.

Len hit a 12½-pounder the next day at 1720 hours and boated a 11½-pound striper at 1740 hours a day later. That ended his hottest streak of the entire year. Seven stripers in his boat in five successive trips that included at least one striper each trip.

"I work a half dozen or more places. I let a place rest one day and come back a day or two later. I vary my trolling speed. If I don't score at one speed and my depth finder shows fish beneath, then I go a little faster and a little slower. Sometimes it works, sometimes it doesn't."

Len also has boated some big blacks and the biggest sand bass of his career trolling the magnum Hellbender. He thinks the bigger fish are deep and that getting down with the deep running lure explains his luck.

I got another call from Len in November. It was a little different from some of the rest.

"What's the limit on Hellbenders?" he inquired.

"Why?" I countered.

"I didn't catch a striper today, but I did hang three magnum Hellbenders, one slab and one jig. That ought to be some kind of a record!"

12

Fishing Santee-Cooper

I pulled the collar of my windbreaker up and slumped as far down in the boat as I could get.

"Slow this rig down," I yelled, looking back toward the stern of the boat where Booty was operating the Merc 20 that was sending us skittering across Lake Marion.

"Is it a little chilly up there?" Booty yelled back.

"That's a good word, but not good enough. It's plain cold," I shouted. "Besides, you don't know this lake and you might hit a stump."

"Not in this one, but maybe in Moultrie," Booty countered.

"Oh yeah! I think you have your lakes mixed. Look at those trees sticking out of the water ahead. Marion has the underwater stumps, Moultrie has the clean bottom."

"Maybe you're right," said my son-in-law. "I sure don't want to put a hole in our boat."

He cut the throttle back and we looked for some of the telltale gulls that would point out where the stripers probably would be feeding. My son-in-law lives in Savannah, and we got the day off to go fishing, by bringing our wives up and turning them loose in Charleston. We took Highway 52 out of Charleston to St. Stephens, then cut over on 45 to the Diversion Canal that links Lakes Moultrie on the south and Marion on the north.

We had spent some time visiting with guides to find out just where to go and how to spot the stripers. We were in the middle of the schooling season on the stripers here. It generally kicks off in early October and runs through February or into March.

I was on a pre-spring break and had waited until now to try my luck. I had been told that chances of getting bigger stripers were better after the first of the year.

I was thrilled just to be here. Santee-Cooper! To a Texas fisherman, Santee-Cooper was the Alamo of striper fishing. This is where it all began!

Booty and I were both scanning the horizon looking for birds. We were strangely alone on the water at this early dawn hour. Back in the east, the promise of a sunshiny day was in the offing. The pinkness that had been on the horizon when we left the dock was turning brighter. The sun was just minutes away from making a spectacular entrance. There is something about being on the water at sunrise that is absolutely breathtaking. I always had been that way in Texas and I had the same sensation here on this storied South Carolina lake.

The guides had suggested such local lures as Striper Swiper or the Striper Strike. If we hadn't found anything by noon, I decided I'd pick up a couple then. But right now we were sticking with some of our favorites, the shallow running Bayou Boogie, Bomber Slabs and the magnum Hellbender. In fact, we had some of the clear plastic Hellraisers with a spinner aft if we actually found the stripers on the surface.

We had been told that at this time of the year the stripers come out of the deepwater at intervals to feed. They said if you spot the gulls, you should catch them near the surface.

The distant whine of another outboard on the port side of our 14-foot Customcraft fiberglass boat had us looking to the south. I saw the gulls even before I spotted the boat. They were about halfway between us and the other boat. And as if their appearance was the signal, from out of nowhere, other boats raced to the spot.

Booty tied on a pearl Boogie and I tried a jighead with a white grubworm. I just took the eight-inch worm off some of the worm rigs I had in my tackle box. I subbed the white grubworm, which is about 2½ inches long, for the other worms.

I could hear the excited whoops of other anglers as they

There they go! When gulls hit the water the boaters take off for the surfacing stripers.

tied on to a striper. Their voices carried over the water as clear as a bell.

"These dern fish got lockjaw," said Booty, after his fourth or fifth cast had failed.

I had the same problem.

A boat on our right already had three stripers aboard. I yelled over to see what they were using. One of the men in the boat held up a white jig with a bucktail.

"Ya don't jerk it or jig fish it, but reel it in smoothly," he said. "When you cast, count to five after it hits the water, then reel in. If you don't score, try six or seven. You just have to locate where they are."

Booty had the Boogie off and had put on the jig and worm as I was talking to the man.

Z-z-z-i-i-i-n-n-n-n-g-g-g-g-g!

"Got 'em!" Booty yelled. And as an old black bass fisherman is likely to do, he whipped his rod skyward to set the hook. He almost fell out of the boat when his line went limp.

I grinned. One thing I had learned years ago. The big stripers have enough pull that when they hit, they set the hook. You don't set it for them. If you do, you get the same result that Booty got. You lose 'em.

The sun comes up and the stripers go down. That's when your fish locators come in handy.

"What happened?" Booty asked. I told him.

Z-i-i-i-n-n-n-n-g-g-g-g! It was my line taking off this time.

I had tried five, six and seven and then started up at four. As I kept a tight line, I told Booty to try four this trip.

My striper came in at about seven pounds.

That was the size of most of the schooling stripers that we hit. We got three, Booty catching up and then going one ahead, out of that school. But it already had made our day. I had been kidding Booty about his Savannah fishing, since all he had been getting was three or four hundred crabs instead of fish.

The gulls were deserting us. Already some of the fishermen were cranking up their motors and starting after the birds.

I yelled over my thanks to the fellows in the neighboring boat for their tips that got us into the stripers.

"Where you all from?" I asked.

"Nebraska!" the older man replied. "I'm visiting my son who lives in Charleston. I come up here two-three times a year. I'm going to retire here, too. We don't have fishing like this where I come from."

Booty kicked on our Merc and were were moseying along watching for gulls and keeping tabs on the other boats. One thing I noticed was that everyone was wearing life jackets. But with all of the underwater stumps, not only was wearing life jackets essential, but so was a pretty slow speed.

"Gulls ahead!"

The sound waffed back to us from a boat half-hidden in the tree skeletons. Almost magically, the whine of the revved-up motors took over. From time to time the whines were punctuated by a s-l-u-u-r-r-p . . . . s-l-u-u-r-r-p sound. That came when the prop hit one of the underwater hazards and kicked the motor up. I was glad we had a Merc with its kick-latch that would flip the motor up and avoid shearing a pin.

We added another pair of stripers, this time in the 10-pound class, before the gulls disappeared. As most of the boats took off, I suggested to Booty that we stay put.

"I've fished enough for stripers to know that they all don't disappear when the gulls do. Mostly, it is the boats rushing up that sends the shad back down and the gulls on their way. Let's stick around and see what we can find."

Booty kicked on his Hummingbird locator. He scooted the boat in a tacking fashion. "Look at that drop-off," he said. "We're right on top of a ridge."

"It could be a real striper alley. Take a run down it and see if there are any fish around," I suggested.

"Look at the bleeps!" he said excitedly. "Those gotta be stripers!"

Six jigs later we hadn't put a fish aboard. We had cast the jigs out, let them sink to the bottom, and promptly got hung up. Even when we tried counting, we had the same luck.

"How deep did you say those magnum Hellbenders would run?" Booty asked.

"Thirty-five to 40 feet," I said.

"We'll, we're in 50-60 foot water, why don't we try trolling?" Booty countered.

The guides all admitted that Lake Moultrie, with its slick bottom, was the best for trolling without interference. But by the same token, if you got into what they called a "structure area" but what we called back in our part of the country a "striper alley" we could be all right.

It was really uncanny. Booty had picked up the stripers on the locator. We had out about 75 yards of line behind the boat. I had a 15-pound test mono line on my Ambassadeur, Booty had 20-pound test. I got the first two fish.

"What am I doing wrong?" he asked.

"Nothing that a 15-pound test line won't cure," I replied.

Booty put down his Ambassadeur and picked up an openfaced Mitchell with 12-pound test line. He put on a silver magnum Hellbender with a green back.

"Get ready," he said. He was scanning the locator and said we had just gone over a huge school.

There was a high pitched z-i-n-g as the striper hit Booty's plug and took off. He quickly cut the motor and I reeled in out of his way. The big striper boiled to the top with Booty cranking furiously to keep up and keep the pressure on.

"Gawd. That's a hog!" I yelled.

I had the landing net, but that monster striper wasn't ready for that yet. This time it made another run away from the boat and headed for the bottom.

"I . . . can't . . . turn . . . him!" Booty gasped.

"Raise your rod tip straight up and hold it," I suggested.

Up it went, but the pressure on the line bent the end almost on top of Booty's knuckles. Then the striper headed back for the surface. The rod straighten to just a half moon. Booty was working the handle furiously.

Now the striper headed right for the boat. And it was coming up all the time. It was less than 25 feet from the boat when it hit the top and rolled in that gar-like fashion aimed at getting rid of those sharp barbs that bit into its lip.

There was no mistaking its size this time. It was a monster.

"Steady, Booty," I said softly. "Bring him on up."

The big fish had been on 20 or more minutes. It was 50-degree weather, but sweat was popping out on Booty's forehead. I had the net in my hand and was learning out as the fish made its first pass by the boat, just out of reach.

And the sight of the boat sent it heading for the bottom once more. But it didn't have the pep it had earlier. Booty kept the pressure on and turned it back toward the top. This time it was within my net range.

"Don't lose it! For gosh sakes, don't lose it!" Booty whispered hoarsely.

I had to run my left hand right down next to the net to get enough leverage to lift the big fish out of the water. It covered half the bottom of the boat when we got it in.

"It's 30 pounds if it's an ounce," I told Booty.

On a scales some time later, it was two ounces over 27 pounds.

The Santee-Cooper Reservoirs have more of a flat bottom that the lakes that were created in my area when dams backed up water over hilly country. As a result, the stripers seek out holes in the bottom of the lake to rest, coming out to feed. And it also means that they are more likely to be suspended vertically during the winter than at other times of the year.

The schooling habits of the sandies are almost identical with the sandies. They are found more often in the early and late part of the day than in the middle. When you hit an overcast day, your chances are better that they will linger longer. And if there is a nice chop on the water, they are more likely to hit than on a day when the lake is a sheet of glass.

When the spawning season at Santee-Cooper ends late in April, generally the herring leads the way back into deeper water. And since this is the prime food for the stripers, they take out in pursuit of their daily bread. This creates a time on the lake that usually lasts until mid-July when live bait pays off.

Oddly enough, shiners are quite effective. But the guides I talked to all said they should be of a large variety, kept lively, and replaced if they die on your hook or quit swimming about. A perfect line for this type fishing would be a 12-pound test. A 4/0 hook should do the trick, but on stripers, make certain that you have one that isn't easily straightened out. Mustad or Eagle Claw are excellent.

One of the guides I talked to recommended drift fishing with the minnows. Especially in Diversion Canal during March and April. Again, surprisingly enough, anglers have been more effective with this method from say 10 a.m. to 2 or 3 p.m. A sinker is needed to get your bait down, but have it 24 to 30 inches above the hook. Some anglers in drift fishing use the slip cork method, adjusting the depth until they score. With three or four rods out at once, it can really be a picnic when you hit a productive hole.

Santee-Cooper has more humps than "alleys" and they all are excellent areas to fish. If you aren't sold on using live bait, you always can fall back on spoons or jigs. The

Bomber Slab and Speedcast Chump are good producers as well as a number of spoons including the Johnson, Hopkins 75, the Topo and a Whopper Stopper spoon with a double V bottom to give it a unique addled shad action.

The Hellbender, Bomber Waterdog, Golden Phantom Golddigger, and Arbogast Mudcat to mention just a few, are quite successful for trolling. The speed of the boat is important here. I have found you need to move out about twice as fast as you do when trolling for blacks or sandies. Again, vary your speed with your results. If you aren't getting a strike at your initial speed, then try a little faster, or slower until you connect.

One last bit of advice for your first visit to Santee-Cooper. In fact, it should be followed wherever you go, even at that favorite fishing hole. Check with the resorts and guides. Find out *what* the fish are hitting today. Find out *where* the fish are hitting today. Find out *how* the fish are hitting today.

If you do that, the chances are pretty doggone good that you'll score tomorrow on what they scored on today.

13

Below the dam

Jiggs Weger lives in a tiny homemade shack about a half mile downstream from the base of the Denison Dam on the Oklahoma side of Red River. His combination wood and tin hut is well hidden in a thicket of willow trees. It is rustic. It is plain. But to Jiggs, it is home.

Jiggs ekes out a living as a commercial fisherman. He fishes with trotlines and he jug fishes. To the two-martini-for-lunch-bunch, Jiggs' domicile might be a hovel. By the standards of most middle Americans it might be no better than a shack. But it is airtight. Dry. And there is one thing about it. When you live in a house like this, you don't have to worry too much about company for dinner.

But lately, Jiggs has been having a lot of "company." Not the two-legged type that make their way overland. But the finny type. Jiggs' house is far enough removed from any direct land route that if a visitor ever made a successful foray through the underbrush, briars and sandburrs by land, they would never return that way again. By water, it is something else. All those trotlines he has been setting out right in front of his hidden hacienda have been producing stripers.

If they don't just eat his bait and straighten out his hooks, they are on. They pose a problem. If he releases

Jiggs Weger at the front door of his river home.

them, they come back again and again. He can't keep them, because the limit is so low. At this writing it is one, but it is about to be raised to three. And it probably will be five or seven or ten before too many years pass.

But even if it was a hundred, it wouldn't help Jiggs.

"You can't sell a dadblamed striper," Jiggs muttered.

Until the stripers took over below the dam, Jiggs had a fine thing going. He was catching plenty of the old flathead catfish in the 30 to 100-pound class. Then there were buffalo and drum. It was a good living for Jiggs. He didn't have to worry about who was President or income taxes. He had no ulcers. He just had a pretty nice living.

I didn't know it at the time, but when I wrote a story about Jiggs Weger and his striper problem for my news-

The path from the river to Jiggs' house.

paper, the *Denison Herald,* I really did old Jiggs a favor. Overnight, the area below the dam became a mecca for striper fishermen. They reasoned if Jiggs could catch them on a trotline, they shouldn't have any trouble at all landing them on their more sophisticated rigs.

A lot of the fishermen, in time, came up with the same question I had often thought about – where did the stripers come from below the dam?

Jack Harper, the genial aquatic biologist who is general manager of the State Fish Hatchery at Durant, Oklahoma, had the most logical answer.

"You have a hydroelectric dam," he said. "When power is generated, it draws water from the big lake and the smaller fish that are sucked into the vacuum created when the generating of power starts, are drawn through the tubes."

Harper said contrary to popular opinion, neither Oklahoma nor any other state as far as he could ascertain, had ever put stripers below a dam.

"We have a big striper population in the Arkansas River network below Keystone in Oklahoma," he said. "In fact, they extend all the way into the state of Arkansas through the series of locks on the river. Those waters below Keystone are navigable through the network of locks. They produce ideal conditions for a natural spawn."

The next question was "why are the stripers below the dam larger than those caught in the lake?" Day in and day out, the stripers brought from below the dam would be from five to ten or more pounds heavier than the catches reported above the dam.

Harper said it was due to the concentration and availability of food. His theory is that every time power is generated, thousands upon thousands of fish of all sizes are drawn through the tubes. By the time they hit the water below the dam, they are addled.

The stripers like deep spots. There is no deeper spot anywhere than where the force of the water close to the dam is discharged. That is a huge hole. Here, not only stripers, but 100-pound catfish and other denizens of the deep make their home.

All they have to do for dinner is just swim lazily up when the water is being discharged and select their menu from the smorgasbord of addled fish that have been on the 100-foot aquatic slide.

Below Keystone, for example, examination of the stomachs of some of the stripers have revealed an occasional crappie. But mostly it is shad, which for reasons known only to stripers, appear to be their prime food source. Harper went so far as to say that shad provided the bulk of the diet of a striper.

"I would reason that the crappie they dined on below the dam came simply because of first, its availability; and secondly, because the striper was getting too fat and too lazy to refuse the offering," Harper explained.

Because all this food was so readily available, it was easy to see that stripers below the dam, even though they came through the tubes as no more than fingerlings, quickly grew in size.

Another interesting report on the lifestyle of the

Emmett Roberts holds up a 24½-pound striper he took out of a pothole below the dam when the water was low.

stripers below the dam came from Harper. Not all of them are roamers. Some stay relatively close to home.

For example, when the water is discharged through the two generating units at the Denison Dam, it winds up making its way down the Red into the Mississippi finally to the Gulf of Mexico. There are many days when there is no power generated. When that happens, the river all but dries up as there is just enough water pushed through to keep the powerhouse operating. Thus, the stripers who have wandered down stream, must quickly find a deep hole or they'll be left high and dry.

They learn this in a hurry. As a result, on impoundments such as this, where the discharged water isn't caught by a lock or another downstream dam, the stripers have a tendency to stay closer to taw.

The situation is completely different at Keystone, where the discharged water enters a series of navigation locks in the entire Arkansas River system. There, the water goes up the Arkansas to Fort Gibson, and then below Fort Gibson back up the Canadian River to below Lake Eufaula. This series of rises and falls, of highs and lows, created by the navigation locks, also creates a perfect spawning condition. As a result, while in 1975 Oklahoma had a statewide limit of just one striper, the limit at Keystone and in the Arkansas River System was raised to three. And as more annual spawns are confirmed, this limit will continue to rise.

There are many ways of catching stripers below the dam. Jiggs Weger used a trotline albeit reluctantly. He caught the stripers in spite of himself, using only cut bait, shrimp, dead minnows and even doughbait.

But the legions of anglers that rushed below the dam and who will continue to show up aren't content with Weger's method. Probably the most popular method was one that made its way from the Carolinas through Tennessee into Arkansas and then into Oklahoma and Texas before going on into the western part of the nation. Call it drift fishing or current fishing. The name isn't important. What is important, it catches fish.

Of all the striper fishermen that I have come to know down through the years, I guess Joe Kreiger must be classified as the expert. Kreiger's striper rig starts with a triple swivel. He ties his line to one eye; a Hellbender on to the second eye with about 18 to 24 inches of leader, and then adds a jig or streamer fly or even a Sputtlebug, a grub-type worm on a jig head to the third eye.

He puts in below Keystone in a boat and gets as close to the dam as the law allows. There is a steel cable stretching across to stop anglers from getting too close. So he ties his boat on the cable, then lowers his triple swivel rig into the water. The speed of the current creates a trolling action on the Hellbender. He lets it out about 50 yards behind his boat. When the stripers swim up to catch the addled shad coming through the tubes, they spot the streamer fly and Hellbender. Wham! More often than not he has two on at once.

Emmett Roberts is another of the more successful below the dam striper fishermen. The short, rather stocky angler likes to fish alone. He uses a Quick 330 open face spinning reel anchored to a nine-foot Eagle Claw Ocean Spinning rod. He has 17-pound test monofilament line. His lure choice varies.

He carries a lot of large, ball jigs painted bright fluorescent colors. He uses some swimming lures, his favorite being the Golden Phantom Golddigger.

"There's two ways to fish for stripers below the dam," he says matter-of-factly. "Then the water is on and when the water is off."

Roberts likes it better when the water is off. He has three potholes a half to three-quarters of a mile below the dam where he says the big stripers hie to in a hurry when the power generating is shut off and the flow of water slows.

He says stripers are smart fish. "They feel it when the water speed slows and if they have gone too far downstream to get back up, they head for a pothole. Since it usually is the bigger fish that makes the longest downstream trip, I always catch my largest stripers when the power is off and the river is just a trickle."

Roberts wades out in the river, wary as a fox for quicksand. Each step is carefully measured and a light pressure put first before he moves forward.

He says there usually is 30 to 40 feet of width to these potholes "and a lot of brush."

Roberts casts toward the brush, halting the flight of his plug in mid-air to try and drop it just short of the brush. He lets it sink to the bottom and then gently raises his rod tip to let the lure dart up and down.

"Sometimes they hit it down in the hole. Then you have a real problem getting them out. You lose a lot that way," he declared. "Others have come up to shallow water to feed and they'll hit the plug close to the top."

Roberts finds early mornings, just after dawn, the best time to fish the river potholes. He often catches his striper and is back in town before there is a bait shop open to weigh in his catch.

One of his biggest, a 24½-pounder, hit right on top. Roberts had seen it surface and with his plug behind it, he flipped the plug toward the top and was skittering it across the top when the big striper hit.

"That was a real funny morning," he said. "Earlier, I had worked a pothole closer to the dam. There was another fellow there. If he hadn't of been there I wouldn't even be telling you this. But I got witnesses so you'll have to believe it.

"These fish were surfacing, I think they were sandies, big goldeneye shad and some blacks. The gulls were diving right on top of the water to get some of the shad the fish were chasing. One gull made his dive and fluttered on the water a second as he selected his bite.

"That's when the gull got bit. I'm not sure whether it was a black bass or a striper. I think it probably was a striper, because it didn't keep coming out of the water the way blacks do. Anyway, it grabbed a mouthful of tail feathers out of that gull. That old gull let out a squawk and took off, and it didn't have a tail feather left. It sure was funny," Roberts laughed.

I was telling that story to Jodie Grigg sometime later. He said he wasn't at all surprised. He recalled seeing a blackbird light on a cattail and the weight of the bird pulled the cattail down until it was about 18 inches from the water.

"Whap! A big lunker black boiled up from the bottom, cleared the water and before the surprised blackbird could get airborne, the bass had it by one wing and was heading for the bottom. The last I saw of that blackbird was its other wing beating the water before it went under.

"You know something," Grigg grinned. "I often thought I ought to design a fishing lure shaped like a blackbird. But I haven't gotten around to it yet."

14

Trolling for stripers

I got a tennis elbow without getting near a tennis court and without hitting a single ball over or into the net. I got my tennis elbow fishing. That's right! I got it fishing.

It all happened when Whopper Stopper Inc. came out with its Magnum Hellbender lure designed especially for getting down 35 to 40 feet trolling for stripers. The first time I went out we caught stripers. Everyone in the boat. That did it. From then on I spent as much time as I could muster trolling for stripers.

To really understand my situation, you'll have to tie one of the magnum Hellbenders on the end of your fishing rod, let out about 60 yards of line, and then hang on while the lure digs down 35 to 40 feet. To get it that deep you have to rev up to about twice the normal trolling speed for blacks. As a matter of fact, when you reach the maximum speed and get the maximum depth, your rod tip has a pretty sizeable arch in it.

The lure has a quiver in its swimming action that reacts like a tiny triphammer on the arm you're using to hold the rod. I've read stories about the deep sea fishermen who battle those saltwater monsters for hours and under the rules of the game, they can't get any help.

It is thoughts along those lines that would race through my mind as I held on to the rod, every muscle in my arm and shoulder aching, waiting for the scream of the reel when the striper hit. When that happened, I was miraculously cured instantly. The pain would vanish and sheer bliss would replace it as the battle to turn the striper before he completely stripped all of the line off the reel.

Then would come the charge toward the boat. Now, instead of holding on, my right hand became a blur as I cranked the reel handle intent on picking up the line as fast as the striper was spewing it, without allowing it to get slack. The first charge usually ended when the striper spotted the boat. He would then slam on his brakes and start back-finning like mad before going into a dive for the bottom. This would give me a chance to catch my breath. I would hold the rod tip high and let him take line out only so fast as the drag would allow. I always set the drag tight enough to allow the striper to take line, but not so tight that the only alternative would be for the line to hold or break.

The next maneuver generally found the striper heading for the surface again, this time 25 to 30 yards out from the boat. He would come right to the top, and unlike the black bass that makes the Rudolf Nuriev ballet-like leap, the striper hits the surface and rolls. The object of the roll is to dislodge the tormenting lure by getting slack through the roll.

But the action here is the same as in bass fishing or any kind of fishing. Keep a tight line. If you're still with it at this point, you're home free if your companion in the boat can use a landing net. You'll get a couple of more short runs, then you can lead the fish right by the boat and into the waiting net.

Sounds simple, doesn't it.

It is simple, if you have the right tackle, the right location, and you find the stripers right at home.

Trolling is without question the most productive method of landing striped bass. It is the one method that all anglers can use. Today, thanks to all of the electronic equipment available, the science of fishing is reduced to being able to read a depth finder and fish locator, usually the one and same piece of equipment.

The stripers hang out on deep water, usually in a state of suspension on one side or the other of what used to be a hill or bluff or pass before the area became inundated when a new lake was built. For those who can recall what the terrain used to be, maybe they don't need a map or

Fishing the downrigger is one way to keep a lure down deep when trolling for stripers. Larry Brown is rigging up the downriggers on his boat.

even a depth finder. But we still recommend one as a sure method of coming up on the right location.

If you have a rather long alley to traverse, make a pass or two over it with your depth finder on, moving at a slow speed so you can read what is beneath the boat. When you locate the fish, toss out a location marker at one end and if it is long enough, at the other.

Next check your equipment. In any case, I use an Ambassadeur 5000 or 6000 with 15-pound test monofilament line. Check your drag to make certain that it is tight, but not so tight that the striper can't peel off line when he makes a run.

Let out about 60 yards of line. This will put you far enough back so that by the time your lure digs down to 35 to 40 feet, it won't be right under the boat. To avoid the tennis elbow that I mentioned at the outset, install rod holders on your boat. That makes the going easy. And don't make the mistake of bringing along that rod with the favorite pistol grip, because there ain't no way you're gonna get that rod handle in a rod holder! I found that out the hard way, too!

If you've made a couple of passes over the area where the fish locator indicated fish, and you haven't had a bobble, vary the speed of your boat. A good rule of thumb on the proper speed is to travel about twice as fast trolling for stripers as you do for blacks. If you haven't done much trolling, then you'll have to use the trial and error method. You see, you need the speed to not only get your lure down to its maximum depth but also to keep it there. You'll often find that at a slower speed, your lure will surface.

Since stripers are an original ocean fish, they seldom seek cover when they are hooked. They'll make a run for the bottom and might even wrap a line around some underwater hazard, but it is more by accident than intent. What a striper will do is take off and run.

Once the fish hits, then cut your motor. If there is more than one line out, the angler with the strike yells first. Then the other anglers start reeling in. And that can really cause some fun, because on more than one occasion, the speedy retrieve made over the spot where the striper hit, can score for the angler coming in. When this occurs, and you have two or more fish on, the problem is keep them on opposite sides of the boat so they won't get all tangled up in each other's lines. The secret to landing them both or all three, should you have trips on, is to bring them in one at a time. And never reel them all the way up to the tip of your rod. Keep them two, three feet or more away so you give them swimming room and allow your netter some room to operate, too.

Let me give you a little advice now that comes direct from Jodie Grigg, who designed the magnum Hellbender. The size of your line will determine the depth you will reach trolling the magnum Hellbender:

30-pound test line at 75 to 80 yards will run at 30 feet.
25-pound test line at 75 to 80 yards will run at 32 feet.
20-pound test line at 75 to 80 yards will run at 35 feet.
15-pound test line at 75 to 80 yards will run at 40 feet.

Since the stripers take off running, you don't need a well rope to land even the biggest of the fish. I remember getting into a running argument with one of the Orvis brothers on occasion of getting my first split bamboo spinning rod and companion Orvis open face reel. I wanted it for use in landing black bass in southwestern lakes generally filled with brush. I suggested a 15-pound test line. I received an indignant reply that four-pound line would be too much and a two-pound line was suggested. Mr. Orvis informed me that he had caught fish in the ocean over 100 pounds on four-pound test line and he wasn't about to believe any old Texas tales of freshwater fish any larger than that.

The point is that blacks, when they strike, head for cover or for something to wrap around, that provides them leverage to free the lure that is painfully sticking them. That's the reason for the heavier line. I am confident that a 15-pound test line will handle any striper that comes along in freshwater.

While the deep trolling is the only way you can catch stripers during certain times of the year, they also can be caught in 12 to 20-foot water trolling. That's when any of the black bass deep runners are ideal baits. If you have a favorite, pick it out and you're in business. Bomber, Waterdog, Hellbender, or whatever lure is popular in your area for getting down deep for blacks. By "deep" at this point, we mean in the 18 to 22-foot range.

Again, the actual process of trolling is the same as when deep. You'll be using small lures with smaller wiggle plates and you'll need the same speed to get them down as you did earlier. You see, the striper is a bigger fish than the black bass and it moves more rapidly through the water. Thus you've got to be moving on out if you hope to attract its attention.

One of the main assets to success in trolling is having a good working knowledge of the lake you're fishing. That's where your Ray Jeff, Lorance, Garcia, or any of the other fish locators come in handy.

If the stripers are deep and you don't have a lure to get down that far, there are ways. There is a lead core line that will give added weight. You can always put a bell weight some six to eight feet in front of your line. All of this will get your lure down to the depth you require. And there is the downrigger.

The habits of the stripers coincide with those of the sandies. In the spring you can troll with shallow running lures such as the Bayou Boogies, Fantail, Swimmin Minnow, Sail Shark, or the Bomber Speed Shad and in five to six-foot water you can score readily. You'll want to work the ledges as the stripers generally will be schooling at about eight to ten feet and will come up to feed.

Then later they'll get down to 15 to 20 feet, and that's

when the black bass trolling lures come in handy. When they get down to 35 and 40 feet, then you have to use that something extra to get down if you are trolling, or the magnum Hellbender.

I would recommend a two-way swivel if you are going to use more than one lure. Here you can use a deep runner coupled with a streamer fly or a white grub worm. They both work like charms. You tie your line to one swivel, then that leaves one for the plug and one for the stream fly or grub worm. And don't be completely startled if you come in with two stripers at the same time if you catch them feeding.

Then there is the straight trolling when you may use a loc-fast snap. This is the only snap swivel I recommend as generally to me a swivel is just something else to go wrong and cause you to lose a fish. And of course, you can simply tie your lure directly to your line. When you are trolling, you are not apt to be changing lures as often as you might be when you are casting and can't find the right combination.

Still another method of trolling, one widely used in coastal areas, is with the downrigger. The advent of the striper in freshwater is creating a demand for the downrigger. It is a most effective way for getting any plug to whatever depth you want to fish.

The downrigger has a weight that is dropped over the back of the boat. You let it out to the required depth, checking the depth gauge as it falls. You must remember that even though it is a heavy weight, once you start trolling at the faster speed, it will not ride directly under the boat. So, to get 35 feet, you must let it out to around 60 feet. It will swing back at an angle, by the combination of the flow of the water and the speed of the boat. But it will get any kind of a lure at whatever depth you require.

When you get a strike, the weight is released and it is on an automatic reel and returns to the rest spot off the stern. You then are completely free to play your fish.

15

Birth of a striper lure

Toledo Bend lake that divides a section of East Texas from western Louisiana was the biggest of the new Southwestern lakes created after the end of World War II. It was one of the later vintage lakes and as a result, the fishing for lunker bass didn't really get good until the late 1960s.

While there are black bass purists who disdain anything except a surface plug to land a black, the size of the lunkers lurking on the bottom of Toledo Bend was enough to get all but the most stubborn of these black bass anglers to resort to the old standby ocean method of trolling.

In fact, there was even a Toledo Bend Black Bass Trolling Club. And about the only lure this club used was the 900 series Hellbender, which had a maximum trolling depth of about 22 feet. But it was landing loads of those five to 10-pound lunkers.

When the weather warmed up, and unless you've ever lived in the piney woods of East Texas you don't really know how hot it can get, those lunkers went deeper seeking a cooler haven in Toledo Bend. And these Toledo Bend bass fishermen were frustrated. They knew the bass were there, but they couldn't reach them. As they were using almost nothing but the Hellbender, they contacted Jodie Grigg, the bait boss at his Sherman, Texas, Whopper

Stopper plant. They told him their problem and asked him if he couldn't come up with a Hellbender that would get down 35 feet or more where they could see those bass hiding out from the bleeps on their fish locators.

Grigg didn't build Whopper Stopper into one of the nation's largest lure manufacturing plants without being practical. It cost a wad of money to design a new lure or change the design on an old one. There are new dies to be made, and they are costly. So he suggested that the Toledo Bend fishermen put a weight in front of their lure to take it down deeper.

Right off he struck out. They didn't want any part of it. And Jodie knew why. You see the weight would destroy the floating feature of the lure. The Hellbender is a floating lure at rest and a deep diver on retrieve. When trolling, if it gets hung up, the chances are that the wiggle plate hit the stump or brush first, rather than the hooks. So the anglers have only to back off, give a little slack, and the lure pops to the surface and they can start again. With the weight, that is lost along with a lot of lures.

To give the Toledo Bend bass anglers what they wanted, it would take the development of a new larger Hellbender with a larger body and a larger wiggle plate.

It was a nice idea all right, but before Whopper Stopper invested heavily Jodie decided to do a little inquiring around. During the months ahead, his travels carried him from Chicago to Louisville to Atlanta, Phoenix, San Francisco. In fact, Jodie pretty well criss-crossed the country visiting tackle markets and trade shows. And everywhere he went he asked questions of fishermen and others about the need for a lure to go down after the deep running fish.

And at the outset, Jodie was quick to admit that mostly he was asking questions about or talking to the bass fishermen.

The response was good. There was a definite need for a lure to get deeper. So by the summer of 1971, a full year after the Toledo Bend trollers had first approached him, Jodie had decided to perhaps give it a whirl. And by then that hot weather had taken over Toledo Bend and the trollers were on his back. At least now he could tell them he was working.

But there was little or no urgency in the project, and the day-to-day press of business didn't give him too much time to spend on it. But he had Earl McCollum and Johnny Spencer, two of his key supervisors, busy whittling

Jodie Grigg checks out the paint job on his magnum Hellbender, which is designed to get deep for stripers.

Here is an assembly line adding the hooks and spinners to Hellbenders that wind up being boxed for delivery by the employees at the end of the table.

out the wooden prototypes that would be the start, and taking a piece of metal and filing down the wiggle plate.

Now you don't just make something that looks like a lure, stick a hunk of metal in its mouth and call it a Hellbender. Or by any other name either. There are routine development phases that must be followed.

First of all while the shape of the body would be like your 900 series Hellbender, it would have a different size diameter and length. And it was the same for the wiggle plate. It required a larger surface area.

And you had to keep in mind at all time that the wiggle plate had to be just the right size to take the larger lure down to the 35 foot depth you were seeking.

As Jodie explained it, "in designing this new Hellbender we had to keep in mind the relationship between the body and the wiggle plate. We needed a body of sufficient size and flotation to counter balance the wiggle plate. If we made the wiggle plate too big, the lure wouldn't run true. If we made it too little, the lure would lose depth."

With that kind of a direct relationship between the size of the body and the surface area of the wiggle plate in mind, the first prototype was completed. By then, it was early in 1972.

And all at once something wonderful happened! Beaver Lake in Arkansas became a hot spot for striped bass. Those same striped bass that had made Santee-Cooper in South Carolina a household word with fishermen. And the 900 series Hellbender was one of the more successful lures. But the fishermen there wanted something to go still deeper. Down about 35 feet.

Echoes from Beaver were heard from Keystone in Oklahoma, and then right in the backyard of Jodie, on Lake Texoma.

Almost overnight there was a desperate urgency to get this new lure on the market, and not so much for the Toledo Bend bass trollers, but for the new legions of fishermen who were going after the biggest freshwater fish of them all, the striped bass.

All of the casual research with the magnum Hellbender disappeared. It suddenly had the No. 1 priority at the huge, sprawling Sherman bait company plant. Stripers, here we come!

Feverishly Grigg, McCollum and Spencer worked over the wooden prototype. It was ready. It was hand-carried to the R&F Tooling Company in Fort Worth, along with the wiggle plate. And now it was the job of the die maker to build a mold for the body and a metal blanking die for the wiggle plate.

It was not simply a matter of duplicating the shape, hook location, etc. of the body when you go from wood to plastic. The designer must make sure that the air chambers in the back section will be ample to allow the lure to float even with the larger wiggle plate. There is quite a bit of engineering required of the die maker to take the wooden body and build the proper die for the plastic body.

Six weeks almost to the day, the dies arrived. And it brought the first disappointment. A miscalculation of a reference point had moved the wiggle plate too far inside. When it was first blanked this was obvious. And it was only borne out when several lures were assembled and checked out.

The body itself was a fine duplication of the wooden prototype, but the combination didn't work. The location of the wiggle plate overcame the counter balance of the body.

So the wiggle plate tooling was scrubbed. New tooling had to be instigated. Nothing could be done but wait. And the pressure continued to mount as more and more fishermen in more and more lakes were demanding a deeper running lure. And word that Whopper Stopper had a magnum Hellbender nearing the market was spreading all over the country.

It was pretty close to seven weeks later that the new wiggle plate blanking die arrived and the nervous crew started assembling the result. It was better, but it still wasn't perfect. This time about 25 percent of the lures were operatable. But still missing was the transfer of floating and balance of the wood to the plastic body.

This resulted in further experimenting. It necessitated altering the shape of the body, making it fatter in some areas, slimmer in others. The hooks were relocated. In effect, the body that first had looked so good had now received a complete face lift. So back to the die maker, this time to make the adjustments. But the changes were too radical to simply alter the old mold. Another new mold had to be made.

This time it was eight more long anxious weeks before the new body die came back. By then, the stripers really were tearing up Lake Texoma right at Jodie's front door. And his phone didn't stop ringing.

At first it looked perfect. The first dozen lures put together worked like a charm. Then a slight malfunction started showing up in about one out of four of the next two dozen tried. Where was it? What was causing it? The three men looked at every facet of the lure. At the connecting link, which for days, they thought was the culprit.

Finally, they found it. The wiggle plate was set .010 of an inch too far in the body of the lure. It was pushing against the dam that divides the head with the hollow body that creates the floation.

That corrected, I got a call from Jodie. "It's ready. Meet me at the Engineer ramp at 4:30 this afternoon and we'll try it."

I was there waiting when Jodie and Earl pulled up in their boat. They had been out since noon running the lures at various speeds checking everything out. They couldn't find any flaws this time.

"Now let's see if it will catch fish," said Jodie matter-of-factly.

During the next 90 minutes the three of us hooked seven large stripers from 10 to 16 pounds. There was no doubt in my mind, the Magnum Hellbender had arrived, and just in time.

On the way into town I asked Jodie if it was time to break the story. "Use it Sunday, but tell them it will be at least 30 days before the lures will make their way to even the local markets," he told me.

But there was still one more disappointment. The connecting link to the lure and the wiggle plate was made through two holes punched in the wiggle plate. Since this was a larger and heavier lure than the 900 series, Grigg found that after 90 minutes to a couple of hours of trolling, the tremendous pressure on the wiggle plate from the connecting link would wear right through the wiggle plate.

"I'm not going to put it on the market until I can solve this," he said.

But the story was out. And if the clamor was a shout before, it became a roar now.

But the violent action of the lure, which was its key fish catching design, was still his problem. After weeks more of experiment, the use of a hardened brass eyelet finally solved the last problem. And 90, not 30 days, after the promised date, the first lure designed to get down to 35 feet for striper fishing was on the market.

Grigg gets reports from lakes over much of the nation. The magnum Hellbender is catching stripers.

In many lakes, it has caught the biggest stripers week after week. He had Eagle Claw designed a 3x extra strong hook for his Hellbender, his Bayou Boogie and his Hellcat series for striper fishing.

"We had some fishermen complaining that the hooks were being straightened out by the fish. I feel that was due more to horsing the fish by the angler than by the fish. But in this business, the customer always is right," Grigg said.

Grigg is not stopping with the Magnum Hellbender. He is continuing to experiment with various other lures that might be adaptable to striper fishing. He is confident that the striper push will continue to grow, and this will be one of the greatest game fish the U.S. has ever seen on freshwater lakes.

He knows that three sizes of his Hellbender will catch them as well as his Boogie, and both the 5/8 and half-ounce Hellcat work beautifully when the stripers surface, as they do on Santee-Cooper and Kentucky Lakes. Before long, as the population grows, they'll be doing that on most lakes.

And there are some side benefits already being felt from the new magnum Hellbender. It has gotten down deep in northern states and Canada and is landing Northern Pike and Muskie.

And the marching and chowder club from the Toledo Bend Black Bass Trolling club are happy, too. They are catching those black bass in hot weather now, and they tell Jodie that one of these days, they might just try their luck on a striper.

16

Jigging for stripers

The first time I ever went jig fishing I was using a nine-foot fly rod with an automatic reel. I had tied a tiny chenele jig on my line and dropped it straight down over a brush pile. I stayed hung up half the time. If I didn't get hung up on the drop, I did after a crappie grabbed the jig and tried to bury itself in the brush. But I also filled a stringer with barndoor crappie.

But jigging for stripers? Now that was something else. But when Larry Brown came onto the dock with an 18-pound striper he said he caught on a jig, he convinced me.

"What do I need?" I asked Brown.

"You just bring your rod this trip," Brown replied. "I'll have the rest."

Now a long time ago I found out that the best way to catch a fish is to go with either a guide or a person knowledgeable about the lake. And when you take a big border lake as large as Lake Texoma, having someone who knows the lake is mighty important.

But what impresses a lot of people more than the success of this guide or that one, is the gear they have on their boat. Larry was no exception. He had that 22-footer rigged up like he was on the Gulf heading for the blue water. It had radio, depth finder, fish locater, downriggers, in fact, there wasn't anything on the market aimed at making fishing easier than he didn't have aboard.

And he started putting it to use the minute we left the dock. By the time we had cleared the harbor, he had talked to a half dozen boats to find out where the fish were hitting.

Fishermen are a funny lot. You meet a guide at the dock and he flops a huge string of fish out of the boat and you ask him what he caught them on or where he caught them, and you might get all kinds of answers. And none of them will be right. But you take the same guide and have him over a spot that is hot right now, and you contact him by radio and, if you're among the "in" group, he'll beam you right into his private fishing hole.

And it was to such a hole that we were heading right now. Larry described it as a high point that juts out underwater with a sharp drop-off on either side.

"These stripers hang on either side, suspended at about 35 to 45 feet in water 60 feet or more deep," he explained.

There were two other boats over the ledge when we neared the area. Larry was busy on the radio and we were close enough to see one of the guides talking into his radio, so we knew that was the fellow who had called us in to the spot.

He was giving Larry directions on how to approach without spooking their fishing and also to put him in an area where he could anchor and we still could fish.

Larry pitched me a pearl Bomber Slab, about a half ounce in size.

"Start with this. We may have to go to the heavier one if it doesn't get down straight enough. Just let it out and sink to the bottom. When it gets down, then come up a couple of turns and work it jig-fashion."

There are a variety of jig-type baits that are popular for this kind of fishing. Speedcast turns out a line of Chump spoons from a third to a full ounce that will get you down in a hurry. The Hopkins 75, the Topo, the old Johnson spoon, and the heavier Dardevle spoons have caught stripers among others, including just the plain old jigs in their many shapes and sizes.

Most of the fish are caught fairly close to the bottom unless they are feeding. Then you get them at most any depth.

You can jig with live bait as well as with hunks of metal. But this is a kind of drift-jigging. You use a bell weight on your line at least 24 to 36 inches above your hook. And your hook must be of excellent quality in a size ranging from 1/0 to 4/0. Let it out until you hit bottom, then reel in at least enough turns to make certain you aren't dragging your bait. About all that would accomplish would be to get hung up.

Since you obviously aren't going to have a flat bottom lake, you need to be jigging your bait gently with your rod tip up. In this manner if the water depth suddenly varies, you won't be caught completely by surprise and you'll be in better control of your bait with less chance of hanging up. You need a little breeze for this kind of jigging, and if you can drift down a striper alley you'll have an excellent chance of scoring.

Jigging lacks the action that you get with trolling. But it has its points. If you hit the stripers at a feeding period, you can get all the action in the world in a hurry. It is not at all uncommon for everyone in the boat to have a striper on simultaneously.

When you are "jig fishing" below the dam you fish in a different fashion. Here, you are now directly over the spot where you are searching for stripers. You have to cast out, allowing the current to carry the jig swiftly downstream. If you score, you pick up a big striper coming up into the swift waters. He may hit going up, but when he turns that's something else indeed.

Most of the jigs used are the heaviest that can be found – or made. It is strange how the stripers have spawned so many new luremakers. "Striper Jigs" are one of the fastest selling striper baits because they will catch fish and you will lose them rapidly.

You can get away jigging with 12 to 15-pound test line in the big pond, but not when you are jigging below the dam. In fact, saltwater tackle is the order of the day for many below the dam anglers. They use those prodigious 13-foot surf casting rods with ball jigs up to three ounces in size, and add a bucktail or a white plastic grub worm.

Each below the dam area has its own unique area for fishing. Below the Kerr Reservoir in the Roanoke River anglers can walk a shoreline ramp to a point that is tantalizingly close to the deep pocket where the big stripers hang out. Even though they whip their three-ounce jigs like bullets with their surf rods, they seldom actually hit their target. But there are sufficient stripers moving about to keep everyone busy. Once the striper hits, then the angler yells out and starts walking or running down the ledge as other anglers raise their rods and give the necessary clearance.

The big job here is in landing the fish and as many get off as are brought in. At Roanoke gaffs are used. At other areas, the fish are played out and all but dragged ashore.

At Texoma, only the handicapped anglers are allowed to move up the wall of the dam to within casting distance of the "striper hole." They can have a helper or walker to take their rod when a strike is made, and the walker races down the wall ledge to the water level to battle the fish. If the handicapped angler can maneuver himself to the same spot, then he'll take over and let the helper use the net.

The more fortunate anglers who have full command of their arms and legs, line the banks and just cast into the water several hundred yards downstream. But they use the same jigs, the same type gear, and often have more success in hooking and landing the stripers.

Below Keystone Dam on the Arkansas River in Oklahoma, the anglers use jigs from boats. There is a cable that runs across the river right below the dam to keep fishermen from getting any closer. So the fishermen put in with boats, tie on to the cable, and whip their heavy jigs as near the base of the dam as they can manage.

When the cable broke one time under the weight of the boats and the power of the discharging water, the Corps didn't look back. This time they replaced the broken cable with a much larger one to prevent a reoccurrence. Fortunately no one was lost or injured when the cable did break, although the temperament of the below the dam fishermen was strained until the new cable was installed.

For some reason, the stripers will hit the silver, white or pearl jigs or spoons or those with a fluorescent color quicker than any other color.

We had a chance to give this a good workout when we reached the striper bank and cast anchor. We were in a school of small stripers, two to five pounds. Almost as fast as we got the jig in the water and down 20 to 30 feet, a striper would hit it. We were releasing them as fast as we brought them up, giving them the chance to grow up into real whopper stripers.

I tried some painted slabs in various stripes and colors and had no luck at all. I shifted back to a pearl Chump and was in business. The fun lasted for about half an hour and then stopped.

Stripers, unlike their kissin' cousin sand bass, are not the greedy type. They have their dining hours and when they occur, you can score repeatedly when you hit a school. This is what makes jigging so popular. Once you find a spot, if you are around at their chow time, you'll really score. I've caught sandies by the hundreds and found them with their mouths and their bellies full of threadfin shad and they still went for the hunk of plastic we offered. You'll catch a striper with food in its stomach, but not in its mouth.

It often is impossible to anchor over a likely area for jigging. When this occurs, you can only drift over the spot. Here is where your electronic equipment comes in handy.

Larry has both a Hummingbird fish locater and a Vexilar Videosonar. The Videosonar scans and charts the bottom and the bottom conditions of the lake, while it locates the fish for you. What this means to jig fishermen is that once the bottom is charted, then some landmarks can be utilized for reaching the area and you don't have to chart it with your Videosonar each trip. And, by being able to see all of the highs and lows on the chart, you get a more reliable picture of what is on the bottom and why fish are there or should be there.

Still on the Vexilar was the chart of the area where we had anchored for jigging. Larry pointed out the drop-off, showed the location of the fish when he had recorded it, and explained that the only thing that will change here is the location of the fish.

"They'll move up and down, particularly the stripers," Larry said. "What we try to do is to hit them when they are moving up to feed."

We kept only our largest striper, one that was a few ounces under five pounds. Larry said the stripers feed pretty close to the major periods on the Solunar table. That was nice to know, but then we've followed the Solunar tables in both fishing and hunting for years.

Jigging for stripers pays a lot of dividends. Sometimes you find them running with schools of black bass and white bass and you score heavily on both. With stripers just catching on across the nation, few of the states had the more liberal South Carolina limits in the mid-1970s. But they were starting to catch up.

Many of the lakes where stripers are being planted do

have smooth bottoms. This makes jigging an ideal way to fish. You have no worries about getting hung. You actually can use a lighter line and smaller jig or spoon or slab. Then if you hit a school of smaller stripers that are feeding, you really can have a ball.

One thing you must watch for in jigging is to keep a close eye on your line for the slightest inkling of a strike. When you get it, you need to set the hook, but not in the back-breaking jerk that you use in trying to tear the head off a black bass. Just whip up your rod tip in a smooth, firm motion. That'll set the barb and the striper will know something is happening. Chances are when this occurs, he'll take off running and the next voice you hear will be the screaming of your reel.

Don't be misled by the smaller stripers. They'll give you a fit after they take off. I've had as much scrapping with a six-pounder as a 16-pounder. But just remember that's a striper not a black bass on. If you want to lose it, try bulldogging it. You just close down your drag so that he can take a little line. Let him run against the drag. When he changes directions, watch out. You'll have a lot of catching up to do.

If you are fishing the windward side, you'll have to let out more line. Your jig won't go straight down. The wind action will carry it out. Keep your finger off the button on your free spool and let plenty of line out. You'll probably be at a good 45-degree angle from your jig. You can pick it up off the bottom and let it sink. By the time you work it under the boat, you'll have picked up the slack. And, if you got a strike earlier, you can be assured you had more line out and you'll have that much more fun in boating the striper.

From the leeward side, have your rod straight out when you let your jig down. Use a 7/8 ounce jig anyway, while you can get away with 5/8 ounce on the windward side. You need to get down quickly here and not under the boat or you'll wind up with a line around the prop or you'll hook another line.

17

Electronic fishing

The red bleeps on the fish locator were coming through loud and clear.

"Those are stripers!" Larry Brown said happily. "Man, if you don't catch a striper here, there's something wrong."

They were down at 40 feet. I had on a 20-pound test line.

"Better hook up the Plath downrigger," Larry said. "With that well rope you got on your reel you might not pull even that magnum Hellbender down that deep."

Plath was the downrigger Larry had on each side of his boat, side mounted on the stern end of the 22-foot craft. Larry checked his Hummingbird for depth, then took a second look at his videosonar. He had charted the area we were in on the graf scanner.

"There's plenty of depth here for you to put the downrigger down about 65 feet," Larry said.

The footage counter was clicking off the feet as it carried the 7½-pound weight quickly toward the bottom. I held it firm at 65 feet, then hooked on my line.

"Take off," I said.

Larry hit the throttle and the boat leaped forward. Since the trolling speed for stripers runs about twice that for blacks, there was a need to get down quickly to 65 feet with the downrigger. As the combination of the flow of water and speed of boat hit the weight, it straighten it out to about a 45-degree angle, which should put my lure at about the 40 feet I was seeking.

"We're coming up on the striper school," Larry sang out. "The boat's over them now, so hang on!"

At that point, if your calculations were right, my lure was some 60 yards out back of the boat. That distance was covered in seconds.

Johnny Johnston, third man in the boat, got the first strike. His reel started singing as the fish hit, and the line flipped free of the downrigger. The weight on the downrigger zipped up on the rewind coil reel, and Johnny was in business.

Larry had cut the engine and Johnny was battling his fish. I already had flipped my line clear of the downrigger and had reeled in and was sitting back watching the battle.

Johnny was getting plenty of encouragement, but he felt it wasn't the kind that he was after. He led the striper up to the boat, saw him sound at the sight of the craft and peel out more line as he dove for the bottom. A couple more passes, a roll or two on top, and Johnny had 'em licked.

"Shall I gaff or net your fish?" I asked.

"Let's just stick with the net so we both will have a chance to bring the fish in," Johnny said dryly.

This was at least a 10-pounder, it might go 12.

"Now that we have some bait, let's go fishing," quipped Larry.

"This is KMA 7721 calling KCT 6610. Come in KCT 6610."

The squawk box on the Citizen Band radio had our number. Larry picked up the mike and pressed a button.

"This is KCT 6610, what's on your mind Johnny boy."

"I had my glasses on you when you hung that striper and I gotta bet on its weight. How much did it go?"

Guide John Smith was doing the querying.

"Something under a 100 pounds," Larry answered.

"When Johnny held it up I guessed 13 pounds. How far did I miss it?"

"We haven't put it on the scales so I'd say you were pretty close," Larry said. "You coming over this way?"

"Not at the moment. We've got a school of big sandies cornered and are having fun for now. We may be around later. Thanks. KMA 7721 signing off."

While all of the conversation was going on, Johnny and I had gotten back in business. We were rigged up and Larry took off again.

He was keeping a close eye on his depth finder to keep us in deep enough water so we wouldn't get hung. He made a wide sweep of the old Red River channel which hit up to 85 feet along here, then straightened out and headed back for the striper alley.

Actually, this was the area where as a youngster I had hiked up and down over the same hills that now provided the terrain that was making this part of Lake Texoma a stripers paradise. And if it was heaven for the stripers, think what it had to be for the fishermen.

This was the storied part of the Red River Valley. We were right near the old Preston Trail, were pioneers drove cattle enroute to the Kansas and Nebraska markets. In this same area, Frank and Jesse James, Belle Star, and John Wesley Hardin, who was born just 25 miles to the east in the Fannin County town of Bonham, used to hide out after robbing a train or a bank.

There were caves whittled and carved by man and time in these sandrock hills. Today, they might be havens for stripers. A century ago they were havens for outlaws and trailriders.

It was hard to believe that today we were skimming over this area across the water of a beautiful blue lake, in a boat so loaded with sophisticated equipment that we could be in direct contact with the sheriff's office in Sherman (and I don't know why I thought of the sheriff at this moment) by just picking up a mike and pushing a button.

When the trailriders set out herding 1,500 to 3,000 cattle to market, from the time they left their ranches until they reached the market and then headed home with the cash, they were out of contact with their home base. In fact, until they rode back up in the corral, they had just disappeared from all contact.

This was hallowed land. Make no mistake about it. But the 20th Century electronic age had changed many things. Electronically we had put a man on the moon in far less time than it took to drive a herd of cattle from Texas to Omaha. And now I could sit in an upholstered seat in a carpeted boat and catch fish so large that even a Texan had to put them on the scales first to keep from bragging too much.

A downrigger on an island lake is still a luxury. But it rapidly becomes a necessity when the only way you can catch stripers trolling is by getting down to them. Most deep runners just won't go that far down. But if you hook up to a downrigger you can reach the depth you need.

Big Jon makes several downriggers that should readily fill a fisherman's bill. There are regular and deluxe models, either of which work happily with stern or side mounts. They have an 18-inch rod which keeps the cannon ball weight – a trademark of Big Jon – far enough astern to keep it out of the motor and out of trouble.

Then there is the electric-manual Big Jon, which has to be the ultimate weapon for the angler. The motor is a 12-volt with a strong gear box and a weatherproof switch. When you go with the motor, you remove the manual handle. And it comes in two sizes.

The Plath downriggers that Larry had equipped on his boat had 18-inch rods. All parts were stainless steel and the 7½-pound weights were hooked to 200 feet of stainless steel wire. A trigger release let the weights down and the footage counter clicked off the depth as the weights fell.

A simple clip held the line off your rod, anchored in a rod holder, and once a fish hit, the clip released the line, the weight was spring-reeled back up, and you had a free line to battle your fish.

On inland lakes the CB channels are used for radio communications. On ocean coastal areas and in the Gulf, the boats are equipped with the Coast Guard frequency. And you really need the radio contact on inland lakes just as much as you do on the coastal waters.

Striped bass are peculiar fish. When they are ready for feeding, they move in quickly. For 30 minutes or more, if you are in that spot, you can score. And since stripers are a relatively new inland freshwater trophy, most states start out with rather low limits. They are inched forward gradually as fisheries are established and the fish grow in size.

Thus, when a fishing guide is out and hits a striper alley that is paying off, he gets busy on his radio and lets all other guides on the lake know just what is going on. They may not rush to his spot, but they probably are close enough to another striper alley that they can quit white bass or black bass fishing and make a run for a striper.

And what did today's fisherman do before the advent

of all of the depth finders? Hummingbird. Vexilar. Ray Jeff. Lorance. Today, there are many excellent electronic companies that make depth finders.

The Vexilar differs in that it puts the results on paper. It records everything on the bottom of the lake, shows the depths and also the fish. While the bottom of the lake and the depth will remain constant, only the fish will change.

The various fish locators or depth finders, that at one time were only in use on the fleet of 800 cruisers that ply their way around Lake Texoma, today are as common on a 15-foot bass boat as a 55-foot Chris Craft.

I caught fish in the muddy Red River years ago on a cane pole and a hook that had a grasshopper or a shiner seined out of a creek with a gunny sack tied between two willow poles. Depth finder? Yeah, I had one of those. It was a slick, pussy willow stick that you pushed in the water ahead of where you wanted to step. More so to check quicksand than water depth in many cases.

And look at what the reel anglers use today. There isn't a major tackle manufacturer who doesn't have a line of free spool reels. Ambassadeur was the first and this Swedish reel captivated American anglers. Today, Penn, Heddon, South Bend, Gladding and all of the tackle companies have a similar kind of reel that makes you wonder how you ever caught a fish in your youth.

But the only fish that I caught in my youth that could be compared to the size of the stripers that you land on freshwater lakes in 30 states today, was mudcats. Oh, I landed my share of 20-30 pounders. I got my name in the paper. They weren't much for photographs in those days. You generally had to shoot somebody to get your picture in the paper.

The first striper I ever saw was the adult planting of stripers in Lake Texoma in 1966. Some went up to 18 pounds. They looked like whales. Today, in Texoma and in hundreds of lakes like Texoma across America from New York to California, fishermen are catching striped bass up to 75 pounds.

And you can't expect to go out and catch a fish that size with a willow pole and a minnow. Sure, they'll hit minnows or cut bait. But in a lake that covers more acres than your eye can traverse, how are you going to know where those fish are if you don't have all of the necessary electronic equipment to pinpoint them for you?

Back in the late 1940s, a fellow I know bought a lot on

the lake and put up a cabin. He was close enough to the water that he could walk down to the water's edge. There was a rather smooth bottom to the lake where he was. He built himself a boat slip for his 14-foot fishing boat. And knowing that fish like cover, he hauled in a lot of brush and dumped it under his boat slip. And then he and his wife would sit on the dock and catch a lot of big barndoor crappie plus an occasional black bass that made its way to their cover.

When word of this got out, my friend was blackballed by all of his fishing associates. You see, he was guilty of what had been going on for generations in coastal areas. The art of chumming. He was providing cover for fish to attract them. That was not considred sporting. Not in a new lake that had so many fish that most could fill a creel on their very first trip.

But my, how things have changed. Today, they have bass tournaments where thousands of dollars in cash are offered and fishermen go out with enough electronic equipment on boats to guarantee that they will at least locate the fish. If they don't catch them, it is not the fault of the equipment they have aboard.

And so it is with stripers. But all of this equipment is necessary if your fishing trip is going to be successful. It just takes the guesswork out of fishing. But once you hook the fish, then it still hasn't changed any of the age old battles of man versus fish. And if the odds continue like they have in boats I have been in, fish will win as often as man.

But Larry had the final word. He had put me on stripers at a time they were hitting. I had an equal to Johnny's first strike aboard. Larry had one two pounds larger.

"What did fishermen do before they invented all of this electronic equipment?" I asked Larry.

Larry looked at me for quite a while. Then he said quietly, "Well, I guess we all just ate chicken instead of fish."

18

Striper tackle

The rectangular plastic box holding the fishing lure was familiar to most Southwestern fishermen. But across the face of the box were several neatly typed words that read "Striper Special."

This happened in the mid-1970s and indicates the tremendous speed with which striped bass fishing had caught hold. The manufacturer couldn't wait for the printer, he typed his own message on the box.

The lure in question was manufactured by Speedcast Inc. of Denison, Texas and was the Golddigger. A little extra weight in the front of the lure carried it deeper and it started catching stripers. At 5/8 ounce it was heavy enough to use below the dam, too.

"I received a call one morning from a jobber who was serving the Platt River in Kansas wanting 40 dozen No. 221 Fantails," said Ramon McVay, the Speedcast bait boss. "It was the time of year when we generally put those lures on the back shelves and speed up production on our spoons."

McVay said he asked the jobber if he was fully awake. "You gotta be kidding," McVay told him. "Do you really want them?"

Assured that he did, McVay checked his stock and found he had 50 dozen and told the jobber.

"Better send me all of them if that's all you have in stock," the jobber told McVay.

"If I can be curious, what do you want this lure for this time of year?" McVay asked.

"Stripers," the jobber replied. "This is the hottest lure right now on the Platt River."

McVay turns out the Chump spoon, a terrific lure for jigging for stripers, plus the Golddigger for 20-22-foot depths and the Fantail for shallow, along with the Topfan for surfacing stripers. Mustad hooks are used in all of his line of baits. And any of his underwater plugs work effectively on the downriggers.

Creek Chub has a Striper Striker, which is one of the more deadly surface lures when the stripers are boiling to the top. The Dardevle spoons from Eppinger are excellent for jigging. Edward A. Eppinger, president of the Dearborn, Michigan firm that has been making excellent fishing lures for more than half a century, said that the line of Dardevle spoons "are just one of several Eppinger lures that can be used successfully for striper fishing."

Dick Kotis, who masterminds the Fred Arbogast Co. in Akron, Ohio, has been making striper lures for a long time. In fact, on the box that packages the Arbogast Scudder is the legend "excellent for stripers." However, when that was put on the box the rather large popping surface lure was designed for saltwater use with special 3/X strong hooks. But it has proven to be just as effective when the freshwater stripers surface.

Kotis said either the Scudder or Dasher, along with the Hammerhead, a deep-runner, and the Mud-Bug have been highly successful on stripers in Santee-Cooper in South Carolina and Percy Priest Reservoir in Tennessee. Kotis said it would take a downrigger to get the Mud-Bug down to the deeper depth. And he admitted that if some of the kingsize stripers got a grip on the Mud-Bug, they conceivably could rip out the hooks.

"But wouldn't that be a whale of an experience," Kotis grinned.

The largest striped bass ever taken on a popper-type plug was landed by Harold Slater in 1964 on a Creek Chub striper strike off the Connecticut coast. The striper weighed in at 66 pounds. That catch more than any other made the surface popping type lures popular during the time of year when the stripers boil to the top of freshwater lakes in their feeding pursuit of the shad.

Both the pistol grip and regulation handle are popular with free spool reels. And you can score with any of the baits on the box.

The Fatso, a 5/8-ounce floating-diving lure, has been effective on stripers at Lake Havasu on the Arizona-California border and actually held the lake record for a time, according to Gary D. Storm, president of the Storm Manufacturing Company that turns out the lures.

In inland lakes where shad provide the principal forage for stripers, Storm was high on the firm's original ThinFin silver shad because it so closely resembles the shape and movement of the shad. And Storm has added extra heavy hardware on the sinking series to the delight of striper anglers.

Rapala caught on to the striper fever quickly too. They have come out with Magnum floating and sinking models in both the floating and sinking lure. The new Magnum size in seven inches in length with 1½-ounce weight on the floater and two ounces on the sinking Rapala. They both come in silver, gold and blue and carry 1/0 treble hooks.

C. S. Turbeville, the president of the Bomber Bait Co. in Gainesville, Texas, frankly admits that there are several lakes in the Striper Country where more stripers are caught on the large Water Dog than on all other lures combined.

Open face spinning rigs land their share of stripers.

The original Bomber, too, has caught its share of the stripers, especially at a 15 to 20-foot depth. Either the Bomber or the Water Dog run effectively when attached to the downrigger.

And when it comes to jigging, the Bomber Slab 8900 series, which is the 1¼-ounce, has been one of the best. It gets down in a hurry and has the action that seems to attract the lurking stripers.

Tulsa, Oklahoma-based Zebco is right in the heart of striper country, not only in the Sooner State, but in the entire United States. And it came out with the Zebco Z-Plug that has been a dandy for stripers. It is a 3/8-ounce lure that comes with a shallow, medium and deep running bill.

Earl's Pearl is Zebco's answer for jigging. It has a jig head and comes with a white nylon tail. You can even add a grub worm on it for below the dam fishing. When the stripers are on top, Zebco's Doll-E-Pop is highly effective for taking the schooling, surfacing fish. And Zebco has been quick to capitalize on the new striper lure market by emphasizing the striper-catching ability of this trio of lures.

In going after the stripers below the dam you need a heavy jig that you can cast a long way, even on saltwater tackle. John Bianco, the ringmaster at the Glen L. Evans Company in Caldwell, Idaho, has a pair of entries in this field with his Lead-Belly and Shadrac Jigs. And don't sell them short as jigging weapons above the dam, either.

Heddon's John M. Marsman calls the Tadpolly "our No. 1 selection for any serious striped bass fisherman." He explains, "The interesting thing about the Tadpolly is that size appears to be important at different times of the year. And all three of our sizes, including the little Tiny Tad, seem to have their place."

Other Heddon lures that have proven effective on stripers are the Torpedo, Chugger Spook and Big Chugg.

Lazy Ike, the Fort Dodge, Kansas, lures, are another old favorite with bass fishermen that have caught their fair share of stripers. In fact, because of the increasing use of downriggers for trolling, there really isn't a swimming lure on the market that won't produce.

Rebel lures are excellent striper bait. They have a stick bait that floats and can be retrieved with the darting action that really sends the surfacing stripers crazy. And the medium and deep running Rebels are equally effective.

As Marsman said, the time of year you're fishing must be considered the rule of thumb for striper fishing. There are times of the year when the stripers are deep and jigging is the only way you can reach them.

Stripers live in a state of suspension, that is, they don't seek holes and curl up like the blacks or catfish. They move up to 35 and 40-foot depths from very deep water and you need a trolling lure to get down after them.

Here the Magnum Hellbender has to be considered one of the best. At this writing it is the only lure known to get down to 40 feet on its own. Other lures need the downrigger.

As the stripers move into shallower water, then plugs such as the Bayou Boogie, Swimming Minnow, Garcia King Spot, Sail Shark, Heddon Crackleback, Arbo-Gaster or Mud-Bug or any of the shallow running lures are extremely effective.

When the stripers are on top, then a surface lure with a darting action like the Rapala, Hellcat, Rebel, Heddon Dying Flutter, along with the popping chugger type surface lures, are hard to heat.

If you have a favorite plug and you know how it works

and at what depth, and if the hooks and the body are such that they'll contain the shock of a 20 to 40-pound or larger striped bass, then you're really in business.

RODS AND REELS

To make those lures work just right, you need the proper rod and reel to handle the larger and deeper running and heavier lures. Right on cue, the nation's tackle manufacturers picked up the ball and are striving mightily to keep the proper rods and reels available.

Garcia calls its Ambassadeur 7000 that was introduced in 1975 "a great new dimension in magnum-power casting reels." It was built to handle gamefish in the 100-pound class – stripers, no less – with a solid aluminum spoon that will hold 350 yards of 25-pound mono. The Garcia-Ambassadeur 6000 takes 275 yards of 15-pound Royal Bonnyl mono, which is the proper weight line to get deep running plugs down to their proper depth.

Garcia has quired the Conlon Rod Company and its line of heavy duty rods with the 5½ to 6-foot fast taper, one piece, heavy butt rods are ideal.

The tackle manufacturers are going to have to quickly adapt to the striper as a national fish. It really is amazing to look back on the Great Lakes Coho flurry of the 1960s and recall all of the emphasis placed by leading manufacturers on Coho lures, Coho rods, Coho reels. In fact, they still have a dominate place in angling catalogues, when it is the mushrooming, fast growing, fast spreading striped bass that is crying for attention.

While fishermen can land a fish as heavy and hefty as the striped bass on almost any equipment, Heddon finds that fishermen have better luck on a rod that is at least six feet long, and preferably 7½ to 8½ feet in the Mark IV or Mark V line.

"I recommend rods with our power plus action and secondarily, those with control flex or vari-power," Heddon's Marsman told me.

In effect, Marsman is saying that with a fish like the striper, you need power in the butt yet enough sensitivity in thc tip to feel what is happening.

Having had some rather hefty stripers on and making runs both away from and then toward me, I can appreciate the tip sensitivity. You must know where the lure and fish are. You can't get it with a thick stick.

Heddon offers a trio of saltwater trolling reels in Models

409, 499 and 445 along with Models 283 and 284 in spinning reels. The latter two are preferable because they have a multiple disk drag system to take the heft and zing of the stripers. They have a smooth, three-race ball bearing system on the crank and main shift coupled with a high speed 4.5 to 1 retrieve.

The Mitchell 308 has a fast 4½ to 1 retrieve and its sister reel the Mitchell 408 an even faster 5½ to 1 retrieve, which makes either much in demand with the striper fishermen who prefer the open bail type fishing.

Penn has a push button reel that ranks high with lots of striper fishermen, primarily because many have become accustomed to the Penn saltwater tackle and are well aware of its durability and quality. And Penn is cutting in heavily in the below the dam equipment when anglers go for the heavy surf rods and reels, many the open face variety.

Fenwick makes a variety of rods in the six to seven-foot category that are favorites with this rapidly increasing breed of striper fishermen. They have the stiff butt and fast taper that still keeps the angler in command when the fish is on.

Lew Childres' line of Speed Sticks can't be overlooked in the rod department either. You can virtually put together the rod of your choice by following the Childres' style, and many striper anglers are doing just that.

Zebco has a complete line of rods and reels in both open and closed faced models. And would you believe that the old original Model 33 Zebco still is quite capable of handling any size striper caught in freshwater, even one that goes all the way up to a plus 75 pounds?

Compac has a fine line of rods and reels. In fact, today there is no reason for an angler ever to be improperly outfitted. Sporting goods stores are the first to pick up a trend in what is required for your area. And they'll have just what you must have if you are going to make a successful forayer of the new super fish.

I've fished with anglers who are gung-ho on open face reels. And they lose few fish. Others have the closed face reels in mind and don't want to change. Shakespeare had a few words for it, and the Shakespeare Corporation has a few rods and reels for the new era of striper fishing, too.

You need a reel that you are comfortable with. Your rod needs strength at the butt, and not a lightning fast taper, but enough of a taper to enable you to retain the sensitivity to know where your striper is at at all times. Remember, this is not a black that will boil to the top and angrily stand on its tail and try and spit the lure in your eye.

The striper sounds more that it surfaces. It makes long runs in many directions, including right at you. When it does come to the top, it rolls in its effort to rid itself of its lure. So the tackle you have must be of the kind that enables you to keep pace with this new fish when you get it on your line.

The striped bass offers the greatest challenge to fishermen in the Seventies, but the challenge can only get larger as the freshwater versions of this fish increase in size, too.

Will Rogers once said "I never met a man I didn't like."

To paraphrase Will Rogers, "I've never used any striper tackle I didn't like."

19

Other end of the line

By Kyra Clift

A few years ago my husband was telling me about an idea for a story he was submitting to a national fishing annual. Since I had been writing a foods column for *All Outdoors Magazine* for several years, I asked him why those kind of annuals always had stories about how and where to catch fish, but never how to prepare them.

"What about the other end of the fishing line?" I asked.

Well, my husband thought that was kind of funny, and he put a postscript on his query with my suggestion. I thought it was terribly nice of the publisher to select my idea for a story even though he didn't go along with the other query.

That's why I insisted on doing my thing, as they say, in this book. I might simply start out and say of all of the various species of fish that I have been called on to prepare in our married life, I must rate the striped bass as the No. 1 fish for eating.

Perhaps because it has family ties with the saltwater fish, it has more of an oily base, which reduces the need for adding butter when simply baking it. The texture of the meat is not unlike lobster, and the flavor is simply delicious. As a result, it would be pretty hard not to prepare a tasty seafood dish when you are starting with such a wonderful fish.

Having gotten this far without any male influence being asserted, I would like to offer as an initial recipe for the striper, one that I have obtained from a member of the male culinary fraternity. Len Andersen comes by his fish cooking abilities honestly as he is of Swedish origin. He not only loves to fish, he loves to cook fish. And when the fish in the Andersen household is prepared outside, Len takes over. Here is one of his favorite ways for preparing striped bass.

SMOKED STRIPER

1 striped bass in the 10-12-pound class
8 pounds charcoal
1 bucket of water
1 sack green mesquite wood
1 pound ice cream salt
1 box table salt

Well now, that's a man's recipe for you, isn't it? The only edible thing so far is the striped bass. But it takes all the rest to make that striped bass come out just right.

First of all, you must have a smoker. Perhaps I should have started out with that. You know the kind. They are about 2½ feet high with a round lid that lifts off. It has ample room at the bottom to hold the charcoal. You put in the charcoal and light it, but without any of the starter fluids that might retain a taste. Start with a little paper and kindling and add a few pieces of charcoal at a time until you get your fire going and add as rapidly as possible until you have all eight pounds burning.

Next, have the mesquite cut in inch-thick slices and completely cover the charcoal. If you can't find green mesquite, then use dry, but soak it in water for at least 48 hours so it won't burn right up. Next fill the container in the pot with water, put the grill in place and bend your fish around it. If the striped bass goes 12 pounds, you probably will have to cut off both the head and tail to get it on the grill.

Count on at least nine hours cooking time. Which means you'll have to get up early to get the fire started and the fish on if it is a working day. At noon, check the water to make sure there is enough to keep moisture in the smoker.

Along about 9 to 9½ hours after you put the fish on, it

This is a smoker, not much to look at, but oh! what it does to a striper.

is ready. You'll find the grill by now just warm to your touch. Carry the fish inside, and using a table knife, gently peel off the skin. On each side of the striped bass you'll find a dark grey line. This is fat and can be easily removed with a teaspoon. There are few bones you can take off the top. If you plan to serve the fish whole, it is ready to go. Put it on a platter, garnish with parsley and thin lemon slices, and serve. It needs no sauce at all.

But smoked striper has far more uses than that. It has been the hit of more than one cocktail party. It flakes off in firm pieces and can be served on your finest silver tray accomplished by assorted crackers. The crackers should be plain, because you don't want to detract from the taste of the smoked fish. It put the smoked shrimp, clam and oysters I had used in the past in the background.

To me, striped bass has a firm meat. And a striper just baked with lemon juice and very little butter, because it has its own natural oils, is one of the most elegant fishes I have ever tasted. I'll have to put it ahead of its saltwater cousin, the red snapper, and its freshwater friend, the rainbow trout.

OVEN BAKED STRIPER STUFFED WITH SEAFOOD

For more different ways of preparing striped bass than Len's smoked method, his wife Siv takes over. This is truthfully a dish fit for a king, whether the Swedish King Olaf or any other reigning monarch. Siv, like her husband Lennert, is a native of Sweden. And like all Scandanavian women – at least all whom I have met – Siv is an excellent cook; makes fantastic breads, bakes native cookies, and of course, excels in preparing fish.

Although Siv holds down a full-time job as a health instructor in public schools, takes an active role in the activities of her two children and her husband, she still finds ample time to prepare what to most America housewives, must seem to be elaborate meals.

For this oven baked striper, you will need a bread pan about two inches deep. Then you'll require the following:

1 striped bass, approximately five pounds
1 bunch green onions, finely chopped
½ pound fresh mushrooms or one equivalent in canned mushrooms
1 can of clams or oysters
1 cup of frozen, deveined shrimp
½ cup crab meat
2 teaspoons parsley
1 teaspoon white pepper
½ stick butter
1 box saltine crackers

Saute the onions in the butter, noting that the Swedish cook prefers real butter to margarine. If fresh mushrooms are used, slice and saute in the butter before adding the balance of the seafood. If you bought a box of soda crackers with four individual packets, take out one packet and roll into crumbs, then soak the crumbs in the juice from the clams or oysters.

Add the onions, mushrooms, seafood, parsley and pepper and salt to taste, to the soaked crackers. If at this point the dressing still appears to be too loose, then add more cracker crumbs. Should it be too hard, then melt some additional butter and add so you get the perfect consistency for a dressing.

Take your striper out of the refrigerator. Wipe it completely dry and then fill its cavity with the dressing. Preheat your own to 375°. Put dabs of butter on the striper and bake for one hour, basting every 10 minutes.

STRIPED BASS CASSEROLE

The next time you plan a buffet dinner for from eight to a dozen people, if you just happen to have a five to six-pound striper in the freezer, this is the only way to go. And one nice thing about it is that it can be completely prepared well in advance, even the day before, so all you have to do is place it in the oven long enough to brown the top.

1 striped bass in the five-pound class, filleted
1½ quarts water
2 tablespoon salt
½ carrot
2 or 3 green onions
3 whole peppercorns
3 whole allspice
½ bay leaf
1 teaspoon sugar
4 tablespoons white wine

Mix all of the ingredients except the striped bass fillets in a big pan and boil for about five to 10 minutes. Then gently place the fillets in the boiling broth. If they are not overly large, about 10 minutes of poaching will suffice. Should the fillets be larger, then give them a little more time to poach. Drain off excess juice and arrange the results in a shallow buttered baking dish. And Siv has these delightful egg-shaped Swedish stainless steel baking dishes that are just right for this particular fish. To prepare the white sauce to go with this fish, you will need:

3 tablespoons butter
1 bunch chopped green onions
4 tablespoons flour
1 8-ounce can of minced clams
1 4-ounce can of mushrooms
3 or 4 fairly good sized Irish potatoes

Saute the green onions in the butter until they are glassy. Add the flour and mix thoroughly before adding any liquids. Add the juice from the can of clams first and then additional broth left over from the fish until you get the sauce medium thick.

Ready for the oven is this striped bass casserole topped with mashed potatoes prepared by Siv Andersen.

Over the top of the fish that you earlier arranged in the baking dish, sprinkle the chopped clams and mushrooms. Cover it with the white sauce.

Meanwhile, the potatoes needed to be peeled and boiled, and then mashed until they are completely smooth. You can either use a spoon or else spritz the potatoes in whatever design you desire around and over the fish. At this point, with everything prepared, all that is left is sprinkling some Parmesan cheese over the potatoes and then placing in an oven pre-heated to 400° until thoroughly heated and brown on top. As a result, if you did all of the preparation either in the morning before your dinner that evening, or even the day before, you have only to cover the prepared dish, place in the refrigerator, and wait some 30 minutes before you would like to have your guests at the table, before putting in the oven.

Served with glazed carrots, a green salad and crisp French bread and you indeed will have your guests demanding more.

FILLET OF STRIPER CHAMPAGNE

Since the advent of striped bass in lakes in the area where we live, I must admit that this has become the "champagne" of all fish for my household. As a result, I am constantly looking for new and unusual ways of preparing it. Since the striped bass is a relatively new species and since the chances of catching a two to three-pounder are far greater than one of the really big ones, I felt that the fillets from such a fish prepared in this fashion should equate the species itself. This is what you'll require:

6 striped bass fillets
2 tablespoons margarine
2 tablespoons grated white onion
1 bay leaf
½ cup chopped fresh mushrooms
2 teaspoons chopped parsley
1 champagne glass of champagne

Put all of the ingredients except the fish fillets into a sauce pan, bring to a boil and then simmer until the contents have been reduced in half. Remove the bay leaf. Add:

2 tablespoons double cream
salt and pepper

Place the striped bass fillets in a buttered baking dish. Add the salt, pepper and cream to the broth, mix well and then pour over the fish. Sprinkle over the top with Parmesan cheese. Bake in an oven pre-heated to 375° for about 25 minutes or until the fish is thoroughly done.

The rest of the champagne kept chilled will go delightfully well with this dish, along with asparagus topped with croutons and sliced hard boiled eggs, a tossed green salad and buttered rolls.

BROILED STRIPED BASS WITH CREAM SAUCE

1 six to eight-pound striped bass
½ pound or two sticks margarine
1½ cups heavy cream
1 tablespoon paprika
¼ cup brandy
1 tablespoon flour
salt and pepper

Again fillet your striper, this time into about six serving pieces. Place in a baking pan, dot with butter and broil, lowering your oven rack so that the fish are not as close to the heat as a steak. Take the rest of the butter, place in a small sauce pan and melt. Add the flour, brandy and cream. Time your sauce so that it will be heated about the time the fish is ready. Pour over the fish just before serving.

WINE STRIPED BASS

I really would be derelict to the striper fishermen in my family if I didn't come up with one good wine recipe with our favorite fish. You see my husband is somewhat of a wine connoisseur along with his piscatorial expertise. So this one is for him. We'll start with:

½ cup corn oil
½ cup white wine
1 teaspoon oregano
1 teaspoon chopped parsley
1 teaspoon freshly ground black pepper
salt
½ lemon, sliced very thinly
1½ pounds striped bass fillets

Cut the fillets into strips about 1½ to 2 inches long and a quarter to a half-inch in width. Mix all of the other ingredients into a fairly deep mixing bowl. Add the striped bass strips, cover the bowl tightly and place in the refrigerator. Do this in the morning or the night before you plan to cook the fish. Stir the striped bass at least a couple of times during their marinating. When ready to cook, dump the fish and the marinade into a pan and bring to a boil, then simmer until the fish is done. Remove from the fire and leave the fish in the heated marinade, serving warm if an entree.

But just about the nicest thing this recipe will do is make a delightful pickup food for a most unusual cocktail snack when served chilled.

20

The last laugh

There was the editor from Dayton, Ohio, visiting. Then there were a couple of other V.I.P.s along whom I wanted to impress. All in all, I had at least a half dozen important people in the party. I had three completely electronically equipped boats. In addition, I had three of the best striper guides in the business with our party. There wasn't a way I figured we could miss.

In the end, there were a couple of extra guests. If you've ever fussed with your wife over seating arrangements at a party when an extra guest suddenly arrives, you will know what I mean. One of the men called and said he was bringing along his son.

"Don't worry about him. I'll have him tow his own little fishing boat," my friend told me.

So I brought along my daughter, Stephanie. When Paul arrived with his boat, it was a 14-foot glass boat with swivel seats and a 9.8 horsepower Mercury motor.

Paul's dad, Harry, was one of our "guides." I am putting the word guide in quotes because while I refer to three of the men in our party as "guides," they are not professional guides. Harry Strief is a developer; Jodie Grigg makes and designs fishing lures, including the lures we were using, and Len Andersen is an engineer at Texas Instruments with a penchant for striper fishing.

Our visiting editor was David Easterly from the *Dayton Journal.* His father, Claud, was editor of the *Denison Herald* at the time. Ben Munson III operated a cotton mill and is a neighbor whom I was especially anxious to see latch on to a striper.

David and Ben went with Jodie, Claud with Len and Harry and I teamed up. That left the two kids together.

Since their boat had no electronic equipment we told them both to just stay with us. "Sort of wallow in our wake, son," Harry told Paul.

But then you know how teenagers mind these days. I guess by now you know what I'm leading up to. I've put everyone in their proper boat, so let me tell you what happened.

We had met at a fishing barge, since the men brought their boats by water to the spot and this was an easy place to board. The fishing barge was located just south of the Engineer Corps launching ramp. Those who trailered their boats – and Jodie and Len did – put in there. Harry and Paul came all the way around the point some 15 water miles from Loe's Highport, where they had their boats berthed.

I got everyone aboard their proper rig, saw to it that the adults had a few cold beers and the kids didn't. Then we shoved off.

There were two striper alleys near our departure site. One runs parallel with the dam, while the other takes off in a northwesterly angle using the intake structure as the southeast point and the tip of an Island off Burns Run as the northwest point.

Both had been producing spectacular catches. The stripers were down suspended in 35-foot water. You could pick them up on your depth finders. They had been hitting the magnum Hellbender. There was no way we could miss. Not with Jodie, Len and Harry along as guides.

Jodie took the lead and headed for the striper alley parallel to the dam. He reached his checkpoint, got a reading on his fish locator and pitched out his marker to identify the south end of the alley.

We had been coming behind at a slower pace, giving Jodie the opportunity to get everything set. Jodie had David and Ben along.

He tied on a pearl-colored Hellbender with a black top on one and used a green and white Hellbender on the other.

Len estimated a distance of some 100 yards behind the first boat and then he got his gang rigged up. Harry and I followed suit the same distance behind Len. The kids were over by the riprap bank and Harry motioned for them to get in behind us.

We worked that spot out real good. Two large sandies taken by Claud and his son, David, in different boats was the only result.

"I've sure found one thing out since I've been striper fishing," Harry told me. "When you fish that deep with that lure, the sandies you catch are the biggest I've ever seen. What's more, I've caught blacks, channel cats, drum and some of just about everything that's in the lake in the deeper water."

I don't know whether I've told you this or not. But when you are holding a rod that has a magnum Hellbender working some 75 yards out behind it in 35-foot water, this is a tremendous pull on your rod. It turns into work in a hurry. I think I got my "tennis elbow" from striper fishing. I sure didn't get it playing tennis. I don't get out often enough. But the muscles in my forearm all the way to my elbow really ache after an afternoon trolling for stripers and holding on to the rod.

But it all is worth while when that striper hits and takes off running with your lure. You don't se the hook. If you do you'll just jerk it out of his mouth. The speed of the boat when the striper hits is enough of a jar to have the fish well-hooked.

You just keep the pressure on. You check your drag to make sure that the striper is stripping out your line at not too rapid a pace. And you keep raising your rod tip in a pumping action, picking up any slack, trying to turn the fish. The technique is completely different from going after the mossback linesides.

By now it became apparent that even though all of our depth gauges showed big bleeps to indicate that there were fish, big fish, down at 35 feet, they weren't having our offering at this time. On about the 10th lap around the striper alley, Jodie leaned over and scooped up his marker and headed for the intake structure.

He didn't need to drop his marker here, because there were two physical markers that you lined up with. Your depth finder told you the rest. The men in Jodie's boat had reeled in when he picked up his marker earlier, now they pitched their lure over the side and let it play out.

The rest of us got in Indian file and followed suit. That is most of the rest of us. The two kids were sort of just playing around in the smaller boat. Rather than coming back to the check point and then heading up the alley in a proper manner, they were still back over by the dam. Harry had a horn on his boat. He hit it twice. The sound finally drifted over the water to where Paul and Stephanie were.

Harry stood up in our boat and motioned in that impatient kind of a hand wave that parents have when they are slightly irritated over how their offspring are behaving.

"If I had him here, I'd whack him with an oar," Harry muttered.

"Don't worry about them, we still are in the same area of the lake and they won't be out of sight," I said.

"I know," Harry said. "But I want him over where we're fishing."

We are trolling too slowly really and my lure surfaced way out behind the boat.

"You're gonna have to either speed up or the stripers are going to have to hit surface plugs if I'm gonna catch a fish," I told Harry.

Harry took one more look over his shoulder in the direction of his son, hit the horn again, and then gently moved the throttle forward so we picked up speed. My lure immediately disappeared below the water and the strong pull was back on the rod tip.

I was looking out the back of the boat, making sure my lure was running properly. Out of the corner of my eye I saw the two kids coming across the lake at an angle that would get them to us about midway on our run up the striper alley.

"Here they come," I told Harry.

Harry looked back. "Yeah," he muttered. "But they'll never catch a striper in the water they're fishing. If you don't mind," he said, still in a slightly irritated manner. "Let them catch up with us and I'm going to tell Paul to get where he belongs so I can tend to our fishing and maybe someone will catch a striper."

It was fine with me. I wanted to reel in my lure and check it anyway. Harry shut off the motor and we drifted in a northerly direction, pushed there by a gentle south wind. We were well off what Harry and the others considered the striper alley area.

The kids were close enough for Harry to start yelling at Paul. That's when it happened.

Z-z-z-z-i-i-i-n-n-n-n-g-g-g-g-g!

It wasn't my reel. Harry was in. I looked quickly for the location of the other two boats. Then I heard Stephanie yell.

"Paul. You gotta fish on!"

Paul was in the back of the boat manning the hand operated engine. He cut the throttle and grabbed for his rod. The boat started going in a circle.

Harry got excited.

"Paul," he yelled. "Shut off the damn motor! Shut if off, Paul, shut off the motor!"

Paul had his rod by now, it had been pulled all the way to the front of the boat as the boat made its swing and the striper now was pulling the boat. He tried to grab the throttle but slipped and fell down in the boat. He almost went over the side.

Stephanie was laughing so hard at Paul's antics she was no help.

Harry was about to have a fit. He was roaring advice at Paul, all of which was ignored because Paul was trying to solve his own immediate problems. If he landed this striper, it wouldn't be by the books. It would be by sheer luck.

"Shut the motor! Hold your rod tip up! Keep your line tight! Don't horse that fish, you'll lose it!"

Harry was tossing one bit of advice on top of another.

Paul had finally regained his original seat at the stern of the boat. He was trying to work the hand rudder with his elbow while hanging on to the fish.

Suddenly the striper headed for the boat.

"Grab the net, Steph!" Paul yelled.

Stephanie got the net but it was tangled up with her rod in the bottom of the boat, where she had put in after reeling in when Paul got his strike. The lure with both treble hooks was firmly caught in the net.

Harry was about to have a stroke. "Dammit Paul! Don't you remember a single thing I taught you!" he roared as the boat still went in a circle and Paul was hanging on to his rod with one hand and trying to keep the boat under control by bending down and catching the hand throttle between his elbow and side while reeling in.

Steph saw that she couldn't get the lure and line out of

the net, so she just got enough line to run to the handle of the net, left the lure there, dropped the rod in the bottom of the boat and stood at ready.

The striper was about five feet down and on its first run, went completely under the boat, before the actual play of the rod turned it.

"Keep your line away from your motor! Pull your motor up or you'll lose the fish! Hold your rod tip up!"

Harry had more tips on his lips that Heloise in two weeks of columns.

Paul and Stephanie both were laughing so hard over the whole thing because by now our two boats were getting close enough that Harry was trying to grab Paul's boat and at least hold it steady.

"Here she comes!" Paul yelled.

Steph made a swipe, the rod jumped up out of the bottom of the boat and into the lake, the net hit the water with a splash.

Harry buried his face in his hands.

"Gawdamighty!" he moaned. "Why did I have to live to see all this!"

"I got it! I got it!"

Steph hadn't stopped when that landing net hit the water. She just kept up a circular motion, the striper swam right into the net, and she picked it up and plopped it in the bottom of the boat.

Harry, unbelievingly, raised up.

Paul had picked up the landing net and was trying to shake the striper out of it.

"P-a-u-u-l!"

There was real anguish in Harry's voice. "Don't lose the damn fish now! Put . . . the . . . net . . . back . . . down . . . in . . . the . . . bottom . . . of . . . the . . . boat. That's . . . a . . . good . . . boy. Now . . . gently . . . get . . . the . . . fish . . . by . . . the . . . mouth! Can . . . you . . . remove . . . the . . . hook! Good!

"Don't hold up the fish that way. You'll lose it again. Put it in your ice chest now!"

Stephanie opened the chest and Paul dumped the fish in on top of the pop in the chest.

They both plopped down.

"Hey, Steph! How about that! I got me my first striper! Ain't that something!"

He got up and took the lid off the box and the striper flopped.

You can see the fish and the plaque any time you drop by the lake home of the Harry Striefs.

Harry yelled. "What are you doing now! Put the lid back on!"

Paul held the lid in one hand and reached down in the box. He pulled out one wet can of Coke and flipped it to Steph, then set one down on the floor of the boat. Then he closed the chest.

"Just getting us a pop. I think we earned it," he said. He took a long pull on the Coke.

"Hey," he said to his dad, who was holding on to the gunwales of the smaller boat as we had finally come alongside. "How many stripers have you all caught?"

"Finish your drink, then get your boat cleaned up," Harry said. "Steph! What's that line doing over the side?"

"Oh, I forgot," replied Steph. "That's my rod and reel."

"Reel it in!" Harry commanded.

"I can't," grinned Steph. "The rod and reel is in the water!"

"Then well-rope it in," Harry hollered. Paul already had started.

"Why don't you two go ahead and fish and leave me and Stephanie alone," he grinned. "We not doing too badly."

At dusk, when the four boats returned to the fishing barge, that striper in Paul's chest was the only fish caught. It had been caught where no stripers exist. It was taken out of boat that had none of the navigable, electronic advantages and it was landed under conditions that were unreal.

Steph and Paul were unusually quiet as everyone climbed out of their boats. Knowing that they had the only striper of the trip, they were slightly embarrassed.

But not Harry. There wasn't a prouder father anywhere at that moment than Harry Strief.

"You should have seen my son, the striper fishermen," he told one and all. "Now that's the way stripers should be caught!" And his fatherly clout of affection across Paul's shoulders almost knocked him into the water.

But this was Harry's moment and he was enjoying it. He had taken his son on his first striper trip and he had scored. And he was the only one who did, too.

If you should happen to be in the Tanglewood Hills area of Lake Texoma some day you can still see that striper that Paul caught. Old Harry had it mounted and it occupies a place of honor on the wall of the Strief lake home. And there is a little bronze plaque under the striper. It tells the whole story.

"Caught by Paul Strief. Netted by Steph Clift."

Epilogue

The Santee and Pinopolis Dams, built by the state of South Carolina in the late 1930s in a hedge against the Depression, did create cheaper hydroelectric power. But it also created the new super fish, the freshwater striped bass.

But then something else happened. Before the dams were built, the flow of water from the Santee and Cooper Rivers into Charleston Harbor was only 72 cubic feet per second. By the mid-1970s, the flow had increased to 16,500 cubic feet per second.

And something funny happens when freshwater and saltwater mix. That water tumbling down the locks below the Pinopolis Dam may look gin clear to the naked eye, but the silt it carries is so tiny it is indiscernable to the eye. And when the freshwater hits the saltwater in Charleston Harbor it gets a positive-negative charge.

The result is that it attracts particles to each other, producing larger particles, until they are heavy enough to settle out. And with freshwater lighter than saltwater, when the tide goes upstream it has the saltwater on the bottom, and when it comes downstream it is on top.

Thus, the problem of getting rid of the shoaling or silting of Charleston Harbor was magnified. Originally, the Army Corps of Engineers and the U.S. Navy had been able to handle the problem. The sludge built up at the rate of

200,000 cubic yards a year. This was removed by dredging, with the removal necessary to keep the harbor open.

The sludge at that rate was dumped into marshy areas off barges manned by the Corps and the Navy. But as time went on, the load got heavier and heavier. Finally it became a major problem.

It is one thing to get rid of 200,000 cubic yards of sludge, but what are you going to do with 10 million yards a year?

"What do you do with 10 million of anything?" asked Jack Lesemann, chief of the Engineering Division of the Corps of Engineers at Charleston.

So it became Lesemann's headache. And a study was directed. Lesemann said the first thing he had to accept was that he had to keep the harbor channel open. To do it, he had to get rid of the build up of 10 million yards of sludge a year.

Just so you can visualize how much sludge that is, if it all were put on a regulation football field each year, it would create a stack of sludge a mile high.

Now Lesemann was well aware of the importance of striped bass. He knew that the Santee-Cooper Reservoir striped bass were the basic fish for stocking lakes and rivers across the nation and the world. Whatever he did, he had to do with the striped bass in mind.

By the middle of 1974 he had his plan. It satisified the authorities, the environmentalists and the ecologists. It kept the striped bass people happy.

From the start, Lesemann knew he had to come up with a solution on how to cut down on the shoaling. His answer is to reduce the flow from its present 16,500 cubic feet per second to 3,000 cfs. While this is well above the original 72 cfs, Lesemann figures that the Corps can live with the 3,000 cfs figure.

To reduce the flow, Lesemann will build another canal from Lake Moultrie, starting at the northeast corner near Russellville and St. Stephens, and empty the water into the Santee River. This will divert 80 percent of the water that now flows into the Charleston Harbor from the Cooper River below the Pinopolis Dam.

With the flow of water reduced by 80 percent on the Cooper River, the number of fish coming up the river to spawn could well be reduced by the same number. And the most important fish is not the stripers here, because they already are in the reservoir and go upstream from the two lakes to spawn. But there is a need to get the blue back herring and other shad that provide the forage fish to come up stream.

When the forage fish famine hit the two lakes in the 1960s, fishery biologists arrived at the happy solution of simply opening the gates of the Pinopolis locks during spawning season to let the spawning-bound forage fish in along with any striped bass. They have been doing that. But with Lesemann's plan diverting 80 percent of the flow, that could be hurt.

But Lesemann thought of that, too. He visited an automatic fish lift operating on the Susquehanna River in Pennsylvania that was designed by the U.S. Fish and Wildlife Service of Massachusetts. That, he figured, would solve his problem of getting the fish upstream during the spawning run.

The fish lift works like this: When the fish start upstream to spawn, because of the heavy flow of water in the canal that will run into the Santee River, the spawning fish will divert to the canal rather than just running up against the dam as they have been doing since 1940 when the dam was closed.

When they hit the various levels on the canal, they'll converge in an area and every 10 minutes a lift will raise them to a higher level until they are able to enter Lake Moultrie. There will be both the forage fish and stripers as well.

In addition, since the bulk of the fish will be coming this way rather than up the Cooper River, a new fish hatchery will be constructed along the canal. This will enable the South Carolina fishery experts to collect the ripe female stripers and strip them of eggs for their hatchery.

Of course, the hatchery on the banks at the Pinopolis Dam will operate there, too.

At this writing in the middle of 1975, five subcontracts with both federal and civilian agencies have been completed on the study, and the solution for eliminating the sludge was down to one of two ways.

It will either be pumped into specially designed barges and hauled out to sea and dumped, or else it will be pumped into a special area, where a pipeline will carry it out to sea where it will be dumped.

The first million dollars has been provided to get the plans and specifications underway. The project has been authorized. Barring continued inflationary costs, it should be held to within $80 million, although it could run higher.

The best educated guess of Lesemann would put the completion no earlier than 1981, possibly a few years later. By the time the project is completed, the first year should bring about a reduction in the flow and disposal of the sludge. However, Lesemann figured it would take 10 years for it to reach its maximum reduction.

But during that decade, and all of the years in front of it, because of the planning done, nothing will interfer with the continued development of the striped bass as America's new freshwater super fish.

Acknowledgements

"South Carolina's Striped Bass Story," Jefferson C. Fuller Jr., South Carolina Wildlife Resources Dept.

South Carolina Wildlife & Marine Resources Department, Dennis Wildlife Center, Jack D. Bayless, Chief.

North Carolina Wildlife Resources Commission, William B. Smith, Assistant Chief, Division of Inland Fisheries.

"Arkansas Striped Bass Project – 1973," Bill Bailey, Special Projects Coordinator, Arkansas Game & Fish Commission.

Arkansas Game & Fish Commission, William E. Keith, chief, Fisheries Division.

"Evaluation of the Striped Bass and White Bass Hybrids After Two Years," R. David Bishop, Tennessee Game & Fish Commission.

"Striped Bass – Hormone Spawning," R. David Bishop, Tennessee Game and Fish Commission.

"Rockfish Egg Introduction and Evaluation in Tennessee," R. David Bishop, Project Leader, Tennessee Game and Fish Commission.

"Striped Bass," Charles Suppes, hatchery biologist, Missouri Department of Conservation.

Georgia Department of Natural Resources, Game and Fish Division, Carl Hall, Regional Supervisor.

Mississippi Game & Fish Commission, Harry Barkley, Fisheries Coordinator.

Oklahoma Game & Fish Division, State Hatchery, Durant, Jack Harper, director striped bass project.

Alabama Department of Conservation and Natural Resources, Game and Fish Division, Barry W. Smith, Assistant Chief, Fisheries Section.

Maine Department of Marine Resources, Lewis N. Flagg, Marine Resources Scientist.

Connecticut Department of Environmental Protection, Cole W. Wilde, chief.

New Hampshire Fish and Game Department, Richard G. Seamans Jr., Chief, Inland & Marine Fisheries Division.

West Virginia Department of Natural Resources, Bernard F. Dowler, District Fish Biologist.

Kansas Forestry, Fish and Game Commission, Roy E. Schoonover, Chief, Fisheries Division.

Colorado Division of Wildlife, Fisheries Research Center, Don T. Weber, Wildlife Researcher.

Tennessee Wildlife Resources Agency, William J. Campbell, Supervisor of Fisheries Research.

Great Lakes Fishery Laboratory, John W. Parsons, Fishery Biologist, Ann Arbor, Mich.

Arizona Game & Fish Department, Alban R. Essbach, Chief, Fisheries Division.

Kentucky Department of Fish and Wildlife Resources, Charles C. Bowers Jr., Director, Division of Fisheries.

Missouri Department of Conservation, Stanley M. Michaelson, Fisheries Biologist.

New York State Department of Environmental Conservation, Eli L. Dietsch, Supervising Aquatic Biologist.

Rhode Island Department of Natural Resources, Division of Fish and Wildlife, John M. Cronan, Chief.

Virginia Commission of Game and Inland Fisheries, William E. Neal, Fish Biologist Supervisor.

Texas Game & Fish Department, Division of Information and Education, W. R. Long, Regional Supervisor.

Nebraska Game and Parks Commission, Bob Thomas, Chief, Aquatic Wildlife Division.

Texas Instruments Ecological Division, Consolidated Edison Project on the Hudson River, TI Ecologist Phil Welsch.

U.S. Engineer Corps, Tulsa District, Tulsa, Okla.

U.S. Engineer Corps, Denison Dam Powerhouse Office, Otis Higgenbotham, Resident Engineer.

"Investigation of Striped Bass Culture in Oklahoma," Jack L. Harper and Ron Jarman, Oklahoma Department of Wildlife Conservation.

"Production of Striped Bass Fingerlings on Oklahoma State Fish Hatcheries," by Ron Jarman and Jack L. Harper, Oklahoma Department of Wildlife Conservation.

"Status and Artifical Reproduction of Striped Bass from Keystone Reservoir, Oklahoma," by Kim E. Erickson, Jack Harper, Gary C. Meninsinger, and Don Hicks, Oklahoma Department of Wildlife Conservation.

Utah Division of Wildlife Resources, Bruce E. May, Project Leader, Lake Powell Fishery Investigations.

Other Cordovan publications for fishing enthusiasts

How to fish for bass

Called "the book for learners," this popular manual on beginning to bass fish is full of basics on tackle, bait and techniques about fishing for bass. It is written in such a way that any beginning fisherman can adapt the information to his chosen style of fishing. It deals with the various kinds of casting, spinning, spin-cast and fly tackle with pointers on how to rig it up keeping rod, reel, line and bait in balance. The learning fisherman also gets a full rundown on how to and where to find and approach "fishing country" on any given body of water. It is a simple, lightly written, conversational primer that the beginning fisherman will profit from. **$3.00.**

Know your fishing boats

Looking for a boat? Here is a concise book about the major points to look for in buying a boat for freshwater, be it for fishing or playing. It includes a full selection of photos of boats from leading manufacturers across the U. S. and how you can get information about them. There are also chapters on handling safety plus tips on powering and maintenance of different types and sizes of hulls. **$2.75.**

Know your fishing baits

Author Jerry Zuber examines artificial baits, probably the thing that has caught more fishermen than fish! Did you know that one of the first thorough books written about artificial baits and how to use them was written by a woman? Nearly 500 years ago? Baits come in an infinite variety and Zuber has touched on most of them in this excellent, basic introduction to a highly complex subject. **$2.50.**

Freshwater fishing in Texas

If you're tired of coming home from fishing trips with nothing heavier than a rusty tackle box, stick a copy of this crisply readable manual in your jeans. Russell Tinsley, author of more than 800 articles and six books on outdoor recreation, gets down to the nitty gritty of freshwater fishing in Texas. Cane pole fishermen and users of more sophisticated tackle alike agree that this is one of the best "how-to" fishing books in print. **$4.75.**

Fishing holes of Texas

This handy fishing reference on the 14 major watersheds of Texas is an excellent gift, but we defy you to buy it without wanting one for yourself. Authored by veteran outdoors writer L. A. Wilke, he has developed an uncomplicated, practical guide to the best fishing spots in the state, complete with lots of maps. It tells you where the fish are and what to expect at each location. Fits a glove compartment or tackle box with ease. **$4.25.**

Fishing the bays of Texas

Just off the press and already looking like an extremely popular seller. Author Husak, who has spent most of his life on the bays, provides details on how to rig up for, seek out and catch the myriad of saltwater fishes of Texas. His style is readable and easy for the novice angler to follow and understand. Covers fishing, fishing spots, camping, cooking and techniques. **$4.75.**

Fishing the Texas Coast – Inshore and Offshore

This saltwater "how-to" book by A. C. Becker Jr. covers everything from surf and wade fishing to pier, jetty and offshore fishing. Gives information on where and how to catch year-round fish like speckled trout, redfish, flounder, gulf trout and others inshore . . . redfish and snapper offshore. Reveals the habits and habitat of migratory fish like Spanish mackerel, tarpon, pompano, jackfish and many more. Gives guidance on baits, care of equipment, how to play certain fish. There's even a section on fish recipes and record catches! **soft cover: $4.50; hard cover: $7.50.**

Panfishing in Texas

If you love to fish, and what you catch is not what you WANT to catch, maybe you haven't yet discovered panfishing. Author Russell Tinsley tells you the panfish varieties available in Texas waters, where they hide and the tackle, rods and reels you'll need to catch them. There's also a discussion on fly fishing versus ultralight spinning. **soft cover: $5.50; hard cover: $8.50.**

Fishing the coast and lakes of Northeast Mexico

Mention Northeast Mexico to a bass fisherman today and odds are he'll immediately respond with some wild tales he's heard about Vicente Guerrero. But did you know you can battle action just as wild with the snook at Rio Soto less than a three-hour drive away? This big four-part book by Stan Slaten gives you the straight facts on the best of Northeast Mexico's salt and freshwater fishing: Guerrero, Rio Soto, Amistad Reservoir, Don Martin, Falcon Reservoir, Azucar Lake, Lower Laguna Madre and others, PLUS details on camping and accommodations, crossing the border and supplies. Over 80 illustrations, index. **soft cover: $5.75; hard cover: $8.75.**

The Texas Fisherman

As the name implies, this monthly publication is by, for and about Texans who fish, both sexes and all ages. The features, photographs and "how-to" content come from the state's top outdoors journalists — everything from catching them to cooking them. It reaches from border to border to bring its readers the latest information plus all the goose-pimply excitement of fishing a new lake or challenging a new kind of fish. You get the where, when, how and what about Texans and their fishing, both saltwater and freshwater.

1 year, $5, 2 years, $8.

Horseman, The Magazine of Western Riding

This monthly magazine, one of the nation's largest equine periodicals, is a lively, informative guide to buying, training, enjoying and winning contests with horses. It features important "how-to" articles by and with the country's leading stock horse trainers plus handy tips on horse production and care from noted breeders. Each issue contains original, educational articles and stories about the handling, exhibition, marketing and history of western horses. For more information on Horseman subscription rates and Horseman Books contact:

Cordovan Corporation

5314 Bingle Road
Houston, Texas 77018